STRUCTURE OF MATTER SERIES

MARIA GOEPPERT MAYER

Advisory Editor

ELEMENTARY THEORY

OF ANGULAR MOMENTUM

ELEMENTARY THEORY
OF ANGULAR MOMENTUM

M. E. ROSE

Chief Physicist
Oak Ridge National Laboratory

New York · JOHN WILEY & SONS, Inc.
London · CHAPMAN & HALL, Ltd.
1957

Library of Congress Catalog Card Number: 57–8893

PRINTED IN THE UNITED STATES OF AMERICA

To Alice

PREFACE

During the past several years the theory of angular momentum has come to occupy a more and more important position in the development of physical theories of nuclear and atomic structure. One reason for this is found in those improvements in experimental techniques whereby it has become possible to measure angular distributions in nuclear reactions, or alternatively angular correlations of successively emitted radiations. From an entirely different quarter we find that the theory, in its modern form, can be very advantageously applied to the formulation and solution of problems associated with the static magnetic and electric nuclear moments which are coupled to the electric and magnetic fields arising from surrounding charges. Here reference is made to problems encountered in low-temperature studies as well as to those met in microwave spectroscopy.

It is therefore hardly surprising that there has been an upsurge of interest in the elements of the theory which are basic for the description of such physical phenomena. The theory of angular momentum is essentially a highly formal one. Its principal ingredients are certain parts of group theory and tensor algebra. In these pages, however, the emphasis is much less abstract than these formidable terms imply. The present work is the result of a course of lectures given at the Oak Ridge National Laboratory in the winter and spring of 1955. In presenting those lectures and in writing these pages it has been my conviction that a clear understanding of the elements of the theory could be helpful to many physicists, and that the ideas as well as the techniques involved can be made available to a majority of those with a graduate-course knowledge of quantum mechanics. There is no implication that one can avoid the use of formalism. The simplification which the present treatment, it is hoped, does achieve is based on two delimiting factors. First, we are concerned here only with the properties of rotations because of their intimate connection with the concept of angular momentum. Second, the reasoning is inductive, and, as the theory initially develops, it makes a "smooth-join" with those aspects of quantum mechanics that are, comparatively speaking, common knowledge. In this way, it is felt that it is possible to make the ideas as well as the analysis transparent and simple. At the same time, this entails little if any essential loss in power and/or elegance in the methods. What is sacrificed is the opportunity to discuss and treat the most general and complicated problems in the most expeditious manner. Actually, this loss of generality is felt in very few places, and these are pointed out in the text. What is gained, it is hoped, is the opportunity to make the ideas underlying the theory and the means of using it available to a much wider audience.

Of course, a number of applications are given as illustrations. No attempt is made to treat all possible applications, nor is any one treated in anything resembling an exhaustive manner. In several cases this would hardly be necessary. For instance, we find that the use of the theory greatly facilitates the calculation of the matrix elements of the quadrupole coupling energy. Once this is done, the determination of the energy levels of the system under consideration is a matter of solving a secular determinant, and further discussion would, therefore, be irrelevant from the point of view of our expressed purpose. Part A is devoted to an exposition of the major elements of the theory. Most, though not all, of the applications are to be found in Part B.

The first chapter is a review of basic principles. Here we present such discussion of operators and their matrix representations as is pertinent for the sequel. In this connection a word of explanation or apology may be in order. To make the discussion more apposite a few references to physical situations are made, and the expression "angular momentum" or "spin" is used. While at this point no formal definitions have been given, it is felt that the reader has some "intuitive" idea of what is meant and would profit by the examples chosen for illustration and the discussion of them. For the purist, these examples would be meaningful on rereading after covering material appearing in Chapter II. Actually, the examples chosen are quite familiar ones, appearing in fairly elementary discussions. The remaining four chapters comprising Part A carry the development through the introduction of the coupling coefficients for vector addition (C-coefficients), the transformation properties of the angular momentum wave functions under rotations of the coordinate axes, irreducible tensors, and Racah coefficients. The applications presented in six chapters of Part B deal with static moments of systems composed of charged particles and elementary magnetic dipoles, particles of intrinsic spin $\frac{1}{2}$ and 1, oriented nuclei (this topic includes angular correlation in cascade disintegrations as well as angular distributions and changes in total cross sections in low-temperature experiments), coupling schemes in nuclear reactions, and wave functions for systems of identical particles.

It is a pleasure to record my gratitude to Drs. L. D. Roberts and J. W. T. Dabbs, for many conversations in which they ably represented the reader's point of view. Dr. A. E. Glassgold rendered helpful service in preparing a first draft of the notes for Part A.

April 1957   M. E. ROSE

CONTENTS

PART A

General Theory

I. REVIEW OF BASIC PRINCIPLES

The background needed for the ensuing chapters is presumably well known to all those who have an acquaintance with quantum mechanics. For ease of reference and for completeness, this chapter provides a brief résumé of the necessary formal constructs.

1. HERMITIAN OPERATORS

To discuss the properties of Hermitian operators we assume that the scalar product of two functions, χ and ζ for example, is defined and has these properties:

$$(\chi, \zeta) = (\zeta, \chi)^* \tag{1.1a}$$

$$(\chi, c\zeta) = c(\chi, \zeta) \tag{1.1b}$$

$$(\chi, \zeta_1 + \zeta_2) = (\chi, \zeta_1) + (\chi, \zeta_2) \tag{1.1c}$$

$$(\chi_1 + \chi_2, \zeta) = (\chi_1, \zeta) + (\chi_2, \zeta) \tag{1.1d}$$

The asterisk indicates complex conjugate; c is a complex number. The detailed form of the scalar product is discussed below.

The Hermitian adjoint Ω^+ of the operator Ω has the property

$$(\Omega^+\chi, \zeta) = (\chi, \Omega\zeta) \tag{1.2}$$

For example, for the gradient operator $\nabla^+ = -\nabla$ if the functions χ and ζ vanish on the boundary of the domain of integration. A Hermitian or self-adjoint operator is its own adjoint. This means that $\Omega = \Omega^+$, and therefore

$$(\Omega\chi, \zeta) = (\chi, \Omega\zeta) \tag{1.3}$$

In the example given above $i\nabla$ is Hermitian.

If we consider the matrix representation of the operator Ω, its Hermitian adjoint is the matrix obtained by transposing (interchanging rows and columns) and taking the complex conjugate of each element. Equation (1.2) may be regarded as the formal expression of this definition since, in conjunction with (1.1a), it shows that $(\Omega^+\chi, \zeta) = (\Omega\zeta, \chi)^*$ or $(\chi, \Omega\zeta) = (\zeta, \Omega^+\chi)^*$. Equation (1.3) is then the formal definition of a

Hermitian matrix. Whether we think of an operator or its matrix representation it follows easily that

$$(\Omega_1\Omega_2\Omega_3 \cdots \Omega_n)^+ = \Omega_n^+ \cdots \Omega_3^+\Omega_2^+\Omega_1^+$$

Two important properties of Hermitian operators are the following:
(a) They have real expectation values [from (1.3) and (1.1a)].

$$(\chi, \Omega\chi) = (\chi, \Omega\chi)^* \tag{1.4}$$

(b) The eigenvalue problem yields real eigenvalues and orthogonal functions. That is, the equation

$$\Omega\psi = \omega\psi$$

with its boundary conditions is satisfied only for certain values of ω, in this case, from (1.4) with $\chi = \psi$, real eigenvalues.[1] The corresponding eigenfunctions are labeled with the eigenvalues ω

$$\Omega\psi_\omega = \omega\psi_\omega \tag{1.5}$$

Consider two linearly independent eigenfunctions ψ_ω and $\psi_{\omega'}$. By linearly independent we mean there exist no non-vanishing constants c and c' such that $c\psi_\omega + c'\psi_{\omega'} = 0$. Otherwise ψ_ω and $\psi_{\omega'}$ would differ only by a multiplicative constant and would be equivalent.[2] From the definition (1.3) of a Hermitian operator it follows that

$$(\psi_{\omega'}, \Omega\psi_\omega) = (\Omega\psi_{\omega'}, \psi_\omega)$$

or

$$(\omega' - \omega)(\psi_{\omega'}, \psi_\omega) = 0$$

If $\omega \neq \omega'$, ψ_ω and $\psi_{\omega'}$ are orthogonal; that is, $(\psi_{\omega'}, \psi_\omega) = 0$. If $\omega = \omega'$, ψ_ω and $\psi_{\omega'}$ may still be orthogonalized since they are linearly independent. Thus, the linear combination $\psi_\omega - (\psi_{\omega'}, \psi_\omega)\psi_{\omega'}$ is orthogonal to $\psi_{\omega'}$ although these two functions have the same eigenvalue—are still degenerate. Actually these eigenfunctions have to be simultaneous eigenfunctions of a whole set of commuting Hermitian operators,[3] and the label ω stands for a set of quantum numbers. Then, in the comparison of independent eigenfunctions of the system, no pair will have identical sets of quantum numbers or eigenvalues. The degeneracy means simply that the eigenvalues of at least one of the operators will coincide. In the usual terminology degeneracy implies equal energy.

[1] We shall restrict ourselves to cases where the eigenvalue spectrum is complete.
[2] If we require that they have the same normalization they would, in fact, differ by only a phase factor (a number of modulus one).
[3] More will be said about the simultaneous diagonalization of a set of commuting Hermitian operators in section 3.

In view of the foregoing we can conclude that the scalar product of two independent eigenfunctions is

$$(\psi_{n'}, \psi_n) = 0 \quad \text{for} \quad n' \neq n$$

Here it is again emphasized that n and n' are collective labels for all the eigenvalues; $n' \neq n$ means that at least one of the eigenvalues is different for the two eigenfunctions. The eigenfunctions can be normalized to unity by multiplication by an appropriate number,

$$(\psi_{n'}, \psi_n) = 1 \quad \text{for} \quad n' = n$$

Thus, the ψ_n are said to constitute an orthonormal set

$$(\psi_{n'}, \psi_n) = \delta_{n'n} \tag{1.6}$$

Furthermore, they are also a complete set, although we shall not prove this here. By completeness we mean there exists no function F such that $(\psi_n, F) = 0$ for all n. As a consequence, a reasonably behaved [4] function f can be expanded in terms of the basis ψ_n:

$$f = \sum_n f_n \psi_n, \qquad f_n = (\psi_n, f) \tag{1.7}$$

The geometrical interpretation of the foregoing is obtained by regarding the set of ψ_n as a set of orthonormal unit vectors in a space of as many dimensions as the set of numbers n represents. This number need not be finite, of course. The expansion given above is the analogue of the representation of "vector" f in terms of its components f_n and the unit vectors ψ_n. Even when the space of n is not finite we are sometimes interested in a subspace which is. Thus, n may represent three numbers n_1, n_2, n_3. Of these, n_1 (the energy, say) has a spectrum with an infinite number (countable or otherwise) of permitted values, n_2 and/or n_3 may represent eigenvalues of Hermitian operators for which the spectrum involves a finite number of permitted values. An example is one of the components of the angular momentum. If n_3 has this property, then the sum (1.7) has a finite number of terms, with n_1 and n_2 fixed. This corresponds to choosing only those "vectors" f that lie in a subspace of the total space.

If the functions under discussion depend only on the space coordinate x, the scalar product in (1.1) is an integral

$$(\chi, \zeta) = \int d^3x \, \chi^*(x) \, \zeta(x)$$

over all values of x. Here d^3x is the volume element. In addition, we will be interested in eigenfunctions which also depend on variables with

---

[4] For those insisting on mathematical rigor a satisfactory definition of "reasonably behaved" is difficult. One can say that the coefficients f_n must exist and that the series (1.7) be summable in some sense, although this is tautological. Nevertheless, it may suffice to say that no difficulties are anticipated in our applications.

discrete ranges. For example, there may be a spinor index s. As an example (which will become clearer later on) the index s assumes $2j+1$ values for a particle of angular momentum [5] j. The wave function can then be written as a column matrix each element of which is a function of the space coordinates,

$$\chi(x) = \begin{bmatrix} \chi^{(1)}(x) \\ \chi^{(2)}(x) \\ \cdot \\ \cdot \\ \cdot \\ \chi^{(2j+1)}(x) \end{bmatrix}$$

and the scalar product involves a sum as well as an integration

$$(\chi, \zeta) = \sum_m \int d^3x \, \chi^{(m)*}(x) \, \zeta^{(m)}(x) = \int d^3x \, \chi^+\zeta$$

In the last equality matrix multiplication of χ^+ into ζ is implied, and here the cross means complex conjugate transpose. Unless otherwise indicated, a scalar product involves summation over all independent variables, which will be an integration over variables with a continuous range and the usual sum for discrete variables.

As already indicated, the sum in the scalar product written above can be considered as the matrix multiplication of the row matrix for χ, obtained by transposing and complex conjugating (i.e., χ^+), times the column matrix for ζ. We can also introduce in this connection the orthonormal set ϕ_m whose elements are all zero except that one in the mth place which is unity. If there are $2j+1$ rows, so that j is a half-integer or an integer, then the ϕ_m look like this:

$$\phi_1 = \begin{bmatrix} 1 \\ 0 \\ 0 \\ \cdot \\ \cdot \\ \cdot \\ \end{bmatrix} \qquad \phi_2 = \begin{bmatrix} 0 \\ 1 \\ 0 \\ 0 \\ \cdot \\ \cdot \\ \end{bmatrix} \qquad \phi_{2j+1} = \begin{bmatrix} \\ \cdot \\ \cdot \\ 0 \\ 0 \\ 1 \end{bmatrix}$$

[5] Angular momentum will be expressed in units of $\hbar$. Thus j and similar symbols are angular momenta divided by $\hbar$. Stated otherwise, we depart from the ordinary cgs units and adopt a new system of mechanical units in which $\hbar$ has the numerical value unity.

Consequently, we can write

$$\chi = \sum_m \chi^{(m)} \phi_m$$

This checks with the foregoing when we realize that the ϕ_m form an orthonormal set. That is,

$$(\phi_m, \phi_n) = \delta_{nm}$$

The case $j = \frac{1}{2}$ is well known from the Pauli treatment of electron spin; here m assumes just two values, which we can take to be $\pm \frac{1}{2}$ for reasons to appear in section 8. Another possibility is intrinsic spin 1, of which the Maxwell field affords an excellent example. A sufficient and complete description for monochromatic fields is given in terms of the vector potential [6] $\mathbf{A}$ or its components which may be represented as follows:

$$A_1 = -\frac{1}{\sqrt{2}} (A_x + iA_y)$$

$$A_0 = A_z$$

$$A_{-1} = \frac{1}{\sqrt{2}} (A_x - iA_y)$$

The reasons for the phases and normalization will be discussed later. The "three-ness" $(2 \cdot 1 + 1)$ of this field arises from its vector character, which makes the intrinsic spin 1; see Chapters V and VII. Of course, as we shall see, the electromagnetic field can have any integer angular momentum ≥ 1, just as an electron can have any half-integer angular momentum $\geq \frac{1}{2}$.

2. UNITARY TRANSFORMATIONS

Of particular importance in the theory of angular momentum are unitary transformations. A unitary transformation is a linear homogeneous transformation which preserves lengths and angles; that is, scalar products are left invariant. The geometrical interpretation is a rotation in the space spanned by the base vectors ψ_n discussed in the last section. That is, we now choose a different basis, φ_n say, and, by completeness of the set ψ_n, the functions φ_n must be expressible as linear combinations

---

[6] We assume that the vector potential is defined with the aid of the Lorentz condition $\nabla \cdot \mathbf{A} + (1/c)(\partial \phi / \partial t) = 0$, where ϕ = scalar potential. Then the scalar potential is defined by $\mathbf{A}$ to within an additive time-independent quantity. The latter is a static field and is of no interest for the dynamical problems with which we are usually concerned.

of the ψ_n as in (1.7). This connection between the two bases (ψ_n and φ_n) constitutes the unitary transformation.

If C is an operator (matrix) which generates the unitary transformation and φ_1 and φ_2 are any two functions or column matrices (so that $C\varphi_1$ and $C\varphi_2$ are the transformed functions), then, by definition,

$$(C\varphi_1, C\varphi_2) = (\varphi_1, \varphi_2) \tag{1.8'}$$

Using the properties (1.2) of the Hermitian adjoint, this is

$$(\varphi_1, C^+C\varphi_2) = (\varphi_1, \varphi_2)$$

Since φ_1 and φ_2 are any two functions, we conclude that

$$C^+C = 1 \tag{1.8}$$

By a theorem of determinants we can write

$$\det C^+ \det C = \det 1 = 1$$

where $\det C$ is the determinant of the matrix C. Since $\det C^+ = (\det C)^*$ it follows that

$$|\det C|^2 = 1$$

But, since $\det C \neq 0$ (that is, it is non-singular), it follows that the matrix of C has an inverse C^{-1} such that $CC^{-1} = C^{-1}C = 1$. Not only can we write the linear equations expressing the new basis in terms of the old, but we can also solve these linear equations and express the old basis in terms of the new. Equation (1.8) is just the relation that defines the inverse [7] of C. That is,

$$C^{-1} = C^+ \tag{1.9}$$

As can be seen from (1.8) or (1.9), the product of any number of unitary matrices (or transformations taken successively) is a unitary matrix (or unitary transformation). Also, if $C = C_1C_2$ and $CC^+ = 1$ and $C_1C_1^+ = 1$, it follows that $1 = CC^+ = C_1C_2C_2^+C_1^+$ or $C_1C_2C_2^+ = C_1$. Since C_1^{-1} exists, we find $C_2C_2^+ = 1$. This is the statement given above with the added result that the inverse of a unitary matrix is also unitary.

The result (1.8) may be written in terms of the matrix elements C_{nm}. These are defined by

$$\psi_n = \sum_m C_{nm}\varphi_m \tag{1.10}$$

where the ψ_n and φ_m are the basis vectors in new and old representations, respectively, and, because of the orthonormality of the φ_n,

$$C_{nm} = (\varphi_m, \psi_n) \tag{1.11}$$

[7] By the definition, if $C = C_1C_2 \cdots C_n$, then $C^{-1} = C_n^{-1} \cdots C_2^{-1}C_1^{-1}$.

Rewriting the operator equations of (1.9), $CC^+ = 1$ and $C^+C = 1$, in matrix notation leads to the unitary properties of the matrix elements C_{mn},

$$\sum_l C_{nl}^* C_{ml} = \delta_{nm} \qquad (1.12)$$

$$\sum_l C_{ln}^* C_{lm} = \delta_{nm} \qquad (1.13)$$

where we have used the fact that $C_{nm}^+ = C_{mn}^*$. Thus the orthogonality property is obtained whether we sum on the row or the column index. One way of looking at this result is to consider the elements in any row, say the nth, as a "vector" $\mathbf{C}_{n-}$; the lth element of $\mathbf{C}_{n-}$ is C_{nl}. Then (1.12) states that these row vectors are orthonormal: $\mathbf{C}_{n-} \cdot \mathbf{C}_{m-} = \delta_{nm}$. Similarly we may consider the elements of any column as constituting a vector $\mathbf{C}_{-n}$, and find from (1.13) that these column vectors are orthonormal, $\mathbf{C}_{-n} \cdot \mathbf{C}_{-m} = \delta_{nm}$.

Once the transformation from one basis to another has been defined, the next question is to determine how the matrix representation of an operator changes when we change the basis. Thus, for an operator Ω, if we know the elements $(\varphi', \Omega\varphi)$, what are the elements [8] $(\psi', \Omega\psi)$? The matrix element of the operator Ω between two states described by wave functions φ' and φ is

$$(\varphi', \Omega\varphi) \equiv (\varphi'|\Omega|\varphi) \qquad (1.14)$$

The transformed functions [9] are $\psi = C\varphi$ and $\psi' = C\varphi'$, and

$$(\varphi', \Omega\varphi) = (C^{-1}\psi', \Omega C^{-1}\psi) = (\psi', (C^{-1})^+ \Omega C^{-1}\psi) = (\psi', \Omega_T\psi)$$

where

$$\Omega_T = (C^{-1})^+ \Omega C^{-1} \qquad (1.15)$$

is the transform of the operator Ω. For a unitary transformation $(C^{-1})^+ = C$ and

$$\Omega_T = C\Omega C^{-1} = C\Omega C^+ \qquad (1.16)$$

That is, the matrix representation of an operator in the ψ basis is Ω_T when its representation in the φ basis is Ω and these are connected by (1.16). The choice of basis is arbitrary, and the one most convenient

---

[8] Sometimes we shall use the symbol $(\psi'|\Omega|\psi)$ for the matrix element $(\psi', \Omega\psi)$. This is especially convenient if the quantum numbers or eigenvalues are used for labeling. Thus, $\Omega_{mn} \equiv (\psi_m, \Omega\psi_n) \equiv (\psi_m|\Omega|\psi_n) \equiv (m|\Omega|n)$.

[9] ψ and φ are a column matrices with elements ψ_n and φ_n, respectively; see equation (1.10).

for calculation is often used.[10] Physical results are, of course, independent of this choice, and the preceding discussion expresses this explicitly insofar as the results of measurements are formulated in terms of matrix elements. For example, C may be the unitary transformation for a rotation of the coordinate system chosen to make the calculation of a physical quantity as simple as possible.

3. DIAGONALIZATION OF OPERATORS

The operator Ω is diagonal in the representation of its eigenfunctions

$$\Omega\psi_n = \omega_n\psi_n \tag{1.17}$$

$$(\psi_m, \Omega\psi_n) = \omega_n\delta_{nm}$$

Let Ω_{mn} be its matrix elements in some other basis φ, so that

$$\Omega_{mn} = (\varphi_m, \Omega\varphi_n)$$

$$\Omega\varphi_n = \sum_m \Omega_{mn}\varphi_m \tag{1.18}$$

Starting in the representation of the φ_n, it is an important problem to determine the unitary transformation C from the set φ_n to the set ψ_n which diagonalizes Ω. For simplicity let us assume that Ω is Hermitian with discrete, non-degenerate eigenvalues. This will be the case in many applications of physical interest. From (1.10) we have

$$\psi_n = \sum_m C_{nm}\varphi_m$$

Operate on both sides with Ω; on the left, use (1.17),

$$\Omega\psi_n = \omega_n\psi_n = \omega_n \sum_m C_{nm}\varphi_m$$

and, on the right, use (1.18),

$$\Omega \sum_m C_{nm}\varphi_m = \sum_m C_{nm} \sum_l \Omega_{lm}\varphi_l$$

Equating these two gives

$$\sum_l \sum_m \Omega_{lm}C_{nm}\varphi_l = \omega_n \sum_l C_{nl}\varphi_l$$

[10] This is completely analogous to the arbitrariness involved in the choice of a coordinate system in the description of any physical process. This does not imply that the description of all physical situations can be given without specifying a coordinate system (e.g., angular distributions) but that the choice is only a matter of convenience. In any event, no matter how we choose the coordinate system, the answer to the problem of angular distributions, e.g., relative to a *physically* defined direction (propagation vector of an incident beam), is always the same.

The coefficient of φ_l on both sides must be the same,

$$\sum_m \Omega_{lm} C_{nm} = \omega_n C_{nl} = \omega_n \sum_m \delta_{lm} C_{nm}$$

The result is a set of homogeneous equations for the matrix elements of the unitary operator C:

$$\sum_m (\Omega_{lm} - \omega_n \delta_{lm}) C_{nm} = 0 \quad \text{for every } n \tag{1.19}$$

The requirement that the determinant of the coefficients vanish, that is,

$$\det (\Omega_{lm} - \omega \delta_{lm}) = 0$$

gives an algebraic equation for the eigenvalues ω_n. Once the eigenvalues are known, (1.19) can be solved for the matrix elements C_{nm} of the unitary transformation and thus the eigenfunctions ψ_n. The homogeneity of the equations (1.19) means that, for each n, all but one of the elements can be determined in terms of any particular one. This element is then fixed by (1.12) or (1.13). The only remaining ambiguity is an overall phase ($e^{i\eta}$ with η real but otherwise arbitrary) for each n. Any convenient convention, say all $\eta = 0$, can be chosen without affecting the answers to any physical result.

A more general problem is the construction of functions which are simultaneous eigenfunctions of several operators. In this connection we shall merely quote this general theorem: The necessary and sufficient condition, that a series of Hermitian operators be diagonalized by the same unitary transformation, is that they commute.[11]

As an illustration of some of the remarks in this section, consider the case of two spins (or angular momenta) $\mathbf{j}_1$ and $\mathbf{j}_2$. If these interact only with an external magnetic field $\mathbf{H}$, the Hamiltonian is

$$\beta \, g_1(\mathbf{H} \cdot \mathbf{j}_1) + \beta \, g_2(\mathbf{H} \cdot \mathbf{j}_2)$$

where β is the Bohr magneton, and g_1 and g_2 are the appropriate gyromagnetic ratios. This Hamiltonian is diagonalized by a function which is a simultaneous eigenfunction of the z-components of each angular momentum operator, i.e., the simple product eigenfunction

$$\varphi_{j_1 j_2, m_1 m_2} = \psi_{j_1 m_1} \psi_{j_2 m_2}$$

m_1 and m_2 are the eigenvalues of j_{1z} and j_{2z}, respectively; see section 5. For example, for spin $\frac{1}{2}$ particles with no orbital angular momentum, these are

$$\psi_{\frac{1}{2}\frac{1}{2}} = \begin{pmatrix} 1 \\ 0 \end{pmatrix} \qquad \psi_{\frac{1}{2}-\frac{1}{2}} = \begin{pmatrix} 0 \\ 1 \end{pmatrix}$$

[11] P. A. M. Dirac, *Quantum Mechanics*, second edition, p. 46, Clarendon Press, Oxford, 1935.

in each space. Now consider an alternative situation. If there is no magnetic field and the angular momenta are coupled in the manner of two magnetic dipoles, the interaction energy is proportional to

$$\mathbf{j}_1 \cdot \mathbf{j}_2$$

But the square of the total angular momentum is

$$\mathbf{j}^2 = (\mathbf{j}_1 + \mathbf{j}_2)^2 = \mathbf{j}_1^2 + \mathbf{j}_2^2 + 2\mathbf{j}_1 \cdot \mathbf{j}_2$$

and, since $\mathbf{j}_1^2$ and $\mathbf{j}_2^2$ must remain diagonal (see Chapter II), the diagonalization of the interaction energy $\mathbf{j}_1 \cdot \mathbf{j}_2$ requires the total angular momentum (i.e., $\mathbf{j}^2$) to be diagonal also. Thus, the eigenfunctions of this Hamiltonian diagonalize the square and z-component of the total angular momentum, with eigenvalues $j(j+1)$ and m, respectively, as well as $\mathbf{j}_1^2$ and $\mathbf{j}_2^2$. These eigenfunctions χ_{jm} are obtained from the $\varphi_{j_1j_2,m_1m_2}$ (which diagonalize j_{1z} and j_{2z}, as well as $\mathbf{j}_1^2$ and $\mathbf{j}_2^2$) by a unitary transformation. The coefficients of this transformation will depend on the six parameters j_1, j_2, j, m_1, m_2, and m, and will be denoted by $C(j_1j_2j; m_1m_2m)$:

$$\chi_{jm} = \sum_{m_1m_2} C(j_1j_2j; m_1m_2m)\, \varphi_{j_1j_2,m_1m_2}$$

The $C(j_1j_2j; m_1m_2m)$ are the Clebsch–Gordan or Wigner coefficients and will be discussed in considerable detail in Chapter III.

4. EXPONENTIAL FORM OF THE UNITARY OPERATORS

Returning to the general discussion, and as preparation for the definition (Section 5) of the angular momentum operators, we consider operators of the form

$$e^a \equiv \sum_{n=0} \frac{1}{n!}\, a^n \tag{1.20}$$

Here a^n is the nth iterated operator: $a^n = a \times a \times a \cdots a$ (n times). The usual rule for multiplication of exponentials holds only if a and b commute.

$$e^a \times e^b = e^{a+b} \quad \text{if} \quad ab - ba = 0 \tag{1.21}$$

Since $+a$ and $-a$ do commute

$$(e^a)^{-1} = e^{-a} \tag{1.22}$$

We make use of this result to establish the following useful theorem: The operator $C = e^{iS}$ is unitary if S is Hermitian. To prove this, take the adjoint of C:

$$C^+ = \sum_n \frac{1}{n!} [(iS)^n]^+ = \sum_n \frac{1}{n!} [(iS)^+]^n = \sum_n \frac{1}{n!} (-iS^+)^n = \sum_n \frac{1}{n!} (-iS)^n$$

Thus,
$$C^+ = e^{-iS} \tag{1.23}$$

From (1.22) we have that
$$C^{-1} = e^{-iS}$$

Therefore, $C^{-1} = C^+$, and C is unitary. The converse of this theorem is also true as can be seen by application of the same reasoning. It also follows that, if $S_1, S_2, \cdots$ are all Hermitian, then the product $e^{iS_1}e^{iS_2} \cdots$ is unitary whether the $S_1, S_2, \cdots$ commute or not.

As an illustration, let us consider the following very elementary example. We discuss a rigid rotator consisting of two masses confined to a plane. The Schrödinger equation is

$$\left(\frac{1}{r^2}\frac{\partial^2}{\partial\varphi^2} + \frac{2M}{\hbar^2}E\right)\psi = 0$$

where M is the reduced mass of the system, r the constant separation of the two masses, and φ the angle made by the line joining them with an arbitrarily fixed line. The wave function is

$$\psi = A\,e^{im\varphi}$$

with A a normalization constant and

$$E = \frac{\hbar^2 m^2}{2Mr^2}$$

In order that ψ be single-valued, m must be an integer.

If the coordinate axes are rotated through an angle α so that the rotator remains in the same plane, the wave function becomes

$$\psi' = e^{-im\alpha}\psi$$

In this case, $S = -\alpha L_z$ is Hermitian, and $C = \exp(-i\alpha L_z)$ is unitary.[12] Actually, the function $\exp(-im\alpha)$ is the matrix representation of C in the basis $\psi_m \sim \exp(im\varphi)$:

$$(\psi_m|C|\psi_{m'}) = \delta_{mm'}\,e^{-im\alpha}$$

when the ψ_m are normalized. Successive rotations through angles α_1, $\alpha_2, \cdots$ are clearly equivalent to a single rotation about the same axis through the angle $\alpha = \alpha_1 + \alpha_2 + \cdots$. Here all the operators $S_1, S_2, \cdots$ commute since they are all proportional to one and the same operator L_z.

---

[12] L_z is the z-component of orbital angular momentum, see section 6: $L_z = (\mathbf{r} \times \mathbf{p})_z = -i(\mathbf{r} \times \nabla)_z$ in our units. If L_z is diagonal with eigenvalue m, $\exp(-i\alpha L_z)$ is diagonal with eigenvalue $\exp(-im\alpha)$.

The very simple example discussed in this section is adequate to illustrate exactly what is meant by a rotation. The function $\psi(xyz)$ may be thought of as a field defining a number to be attached to each point of space. After the rotation, to which we referred, the function ψ becomes a new function $\psi'(xyz)$ such that at each point P the value of ψ' is equal to the value of ψ at a point Q, where Q would be carried into P by the rotation. It is therefore clear that, while we speak of a rotation of the coordinate axes, it is convenient and valid to picture this as a rotation of the points and set of numbers constituting the field and, in fact, we shall sometimes refer to the rotation of the field. In the example just discussed the number representing ψ at Q is rotated from its position at Q to the point P. This is formulated more generally in section 6; see equation (2.9).

An extension of this description is immediate if at each point there is not just one number but a set of numbers. For instance, there may be three numbers which constitute the components of a vector; see section 21. Then after the rotation the three components of the vector at P are linear combinations [13] of the three components which existed at Q before the rotation. We may speak of a rotation of the coordinate axes or, equivalently, of the field. The same result is achieved in the two cases if the rotations are inverse to each other.

[13] The particular form of these linear combinations is related to the *transformation law* for the coordinates of a point in space; see equation (5.40).

II. THE ANGULAR MOMENTUM OPERATORS

5. DEFINITION OF ANGULAR MOMENTUM OPERATORS

From the observational point of view, the physical quantity to which the term angular momentum refers is usually the result of a fairly indirect measurement. This is discussed at some length in section 9. The indirectness of the measurement implies a certain theoretical result whereby the quantity identified with angular momentum is deduced by comparison with the data. As a consequence, we are obliged to recognize that the subject must begin with some conceptual basis for the quantity called angular momentum. In the theory this appears as a quantity related to the eigenvalue of an operator having to do with the rotational properties of the physical system.[1]

This is hardly surprising. It will be recalled that in classical mechanics the orbital angular momentum of a system was a constant of the motion if the classical Hamiltonian was invariant under rotations. This meant that the physical system was rotationally invariant or symmetric. In formal terms this implied that the Poisson bracket of the Hamiltonian and the angular momentum vanished. In quantum mechanics this would correspond to the statement that the Hamiltonian and the angular momentum operators commute.

There is, of course, an important difference in the classical and quantum cases. We actually are obliged to define three angular momentum operators just as one deals with three angular momentum components. This is a consequence of the obvious and trivial-sounding fact that we are concerned with rotations in a three-dimensional space. As will be seen, we may have commutation between the Hamiltonian and all three angular momentum components, but there cannot be commutation between any pair of operators corresponding to a pair of components.

---

[1] Measurement of an angular distribution provides an excellent illustration. The measured distribution is compared with the calculated one, and the latter depends parametrically on the quantity called angular momentum. But this does not explain how this quantity entered in the theory. The fact that it does appear has to do with the description of the quantum-mechanical states of the system and in particular, as stated in the text, with the symmetry properties under rotations.

Hence no more than one component operator can be a constant of the motion.

There is another important difference between the quantum and classical cases. In the classical situation one is concerned only with orbital angular momentum or at least with such angular momenta as are associated with a space–time description. This limitation is too stringent for quantum mechanical systems, and a wider conceptual basis for the angular momentum (permitting the treatment of intrinsic spin) must be introduced. Without pursuing this point further, we simply remark that the wave function of a physical system will contain all the necessary information about it. The definition given below is completely general. It emphasizes the essential connection between angular momentum and transformation properties of a system under rotations. These transformation properties are expressed in terms of what happens to the wave function of the system when the rotation is carried out.

To define the angular momentum operators we generalize the rather trivial situation discussed in section 4. Let us consider a rotation of the coordinate system through an angle θ about an axis defined by the direction $\mathbf{n}$, a unit vector. The wave function ψ in the original system is related to the wave function ψ' in the rotated system by a unitary transformation

$$\psi' \equiv R(\mathbf{n}, \theta)\psi \qquad (2.1)$$

The unitary transformation operator $R(\mathbf{n}, \theta)$ depends, of course, on three angles in the general case—two to define the direction of $\mathbf{n}$ and another, which is θ, to give the magnitude of the rotation.[2] Of course, $R(\mathbf{n}, \theta)$ must approach unity in the limit $\theta \to 0$, and it is convenient to write it in exponential form,

$$R(\mathbf{n}, \theta) = e^{-iS(\mathbf{n},\theta)} \qquad (2.2)$$

with the understanding that $S(\mathbf{n}, \theta) \to 0$ for $\theta \to 0$. More important, since R must be unitary it follows that S must be Hermitian.

It is advantageous to consider infinitesimal unitary transformations, i.e., terms in $S(\mathbf{n}, \theta)$ of order higher than the first power are neglected. To this order we write

$$R(n, \theta) = 1 - iS(\mathbf{n}, \theta)$$

or

$$R\psi = (1 - iS)\psi \qquad (2.3)$$

The angular momentum operators are defined below in terms of the transformation properties of the wave function under rotations of the

[2] The Euler angles will prove a convenient choice for these three angles.

coordinate system. The properties of the angular momentum operators are independent of the magnitude of the angle of rotation. There will thus be no loss of generality in considering infinitesimal rotations such that all field quantities undergo only first-order changes in the parameter θ characterizing the rotation. This device, which we shall use on a number of occasions, greatly simplifies the analysis.

For an infinitesimal rotation, the usual Taylor series expansion can be made for $R\psi$ wherein the difference $R\psi - \psi$ is proportional to θ. Thus, for rotations about the x, y, and z axes, respectively,

$$R\psi - \psi = -i\theta J_x \psi$$

$$R\psi - \psi = -i\theta J_y \psi \tag{2.4}$$

$$R\psi - \psi = -i\theta J_z \psi$$

where J_x, J_y, and J_z are to be determined in each case from the transformation properties and structure of ψ (see below). The three Hermitian operators J_x, J_y, and J_z thus defined are the operators for the three Cartesian components of angular momentum. They form a complete set in that the components formed in any other way are expressible as linear combinations of these three.

For an arbitrary axis of rotation $\mathbf{n}$ this is, more generally,

$$R\psi - \psi = -i(\mathbf{n} \cdot \mathbf{J})\theta \, \psi \tag{2.5}$$

with θ the angle of rotation. Equation (2.3) can also be written

$$R\psi - \psi = -i \, S(\mathbf{n}, \theta) \, \psi \tag{2.6}$$

Comparison of these last two equations shows that

$$S = (\mathbf{n} \cdot \mathbf{J})\theta \tag{2.7}$$

More generally, the angular momentum $\mathbf{J}$ is defined by the equation

$$R\psi = e^{-i\theta(\mathbf{n} \cdot \mathbf{J})} \, \psi \tag{2.8}$$

where $\mathbf{n}$ defines the rotation axis, and θ is the rotation angle. The angular momentum thus determines the transformation properties of a system under rotations of the coordinate system. Inversely, the angular momentum operator $\mathbf{J}$ (that is, three operators) can be determined from the transformation properties of the system. This is the essence of our definition of angular momentum. Its particular form is defined according to (2.7) by the transformation properties of the particular system under discussion.

6. ORBITAL ANGULAR MOMENTUM

An important illustration of the remarks of the preceding section is the case of a single function of the space coordinates $\psi(xyz)$, i.e., a scalar field. For example, $\psi(xyz)$ may be the pressure in a sound field, or the Schrödinger function of an alpha particle. Let the Cartesian coordinates of a point be x, y, z in the original coordinate system and x', y', z' in the rotated system. By the definition of a scalar field, the rotated function $R\,\psi(xyz)$ is the original function of the rotated coordinates $\psi(x'y'z')$. Thus,

$$\psi(x'y'z') = R\,\psi(xyz) \tag{2.9}$$

For an infinitesimal rotation $d\theta$ about the z-axis,

$$x' = x + y\,d\theta$$

$$y' = y - x\,d\theta$$

$$z' = z$$

Expanding the left side of (2.9), we write

$$\psi(x'y'z') = \psi(xyz) + d\theta \left(y\frac{\partial}{\partial x} - x\frac{\partial}{\partial y} \right) \psi(xyz)$$

From (2.5), however, we have for the right side of (2.9)

$$R\,\psi(xyz) = (1 - i\,d\theta\,J_z)\,\psi(xyz)$$

and therefore, in this case,[3]

$$J_z = -i\left(x\frac{\partial}{\partial y} - y\frac{\partial}{\partial x} \right) \equiv L_z \tag{2.10a}$$

L_x and L_y may be obtained by cyclic permutation of the symbols x, y, and z. Thus,

$$L_x = -i\left(y\frac{\partial}{\partial z} - z\frac{\partial}{\partial y} \right) \tag{2.10b}$$

$$L_y = -i\left(z\frac{\partial}{\partial x} - x\frac{\partial}{\partial z} \right) \tag{2.10c}$$

[3] The result (2.10a), and therefore (2.10b) and (2.10c), may be deduced without the use of the infinitesimal rotation using cylindrical coordinates and proceeding as in section 4. Thus, for a rotation around the z-axis the azimuth angle is changed from φ to $\varphi' = \varphi - \alpha$. Hence,

$$R\psi = \sum_{n=0}^{\infty} \frac{(-\alpha)^n}{n!} \frac{\partial^n \psi}{\partial \varphi^n} = e^{-i\alpha L_z}\,\psi$$

This agrees with (2.10a) when it is noted that $L_z = -i\partial/\partial\varphi$.

For the case of a single function of x, y, z, i.e., a scalar field, we have easily obtained the angular momentum operators (2.10) from the transformation properties (2.9) of the field. These are the familiar orbital angular momentum operators. If, however, ψ had had a spinor index, such as mentioned in section 1, there would have been changes (such as transpositions) in the spinor components as well as changes coming from the variation in the configuration space coordinates. As we have just shown, the latter are generated by the orbital angular momentum operators (2.10), which are linear differential operators. The changes in the spinor components are effected by square matrices whose rows and columns are labeled by the spinor index. For example, see section 1,

$$\psi' = R\psi = \begin{bmatrix} 0 & 1 & \cdot & \cdot & \cdot & \cdot \\ 1 & 0 & \cdot & \cdot & \cdot & \cdot \\ \cdot & \cdot & 1 & & & \\ \cdot & \cdot & & 1 & & \\ \cdot & \cdot & & & 1 & \end{bmatrix} \quad \psi = \begin{bmatrix} \psi_2 \\ \psi_1 \\ \psi_3 \\ \cdot \\ \cdot \end{bmatrix}$$

transposes the first two components of ψ. As a consequence we find in general that $\mathbf{J} = \mathbf{L} + \mathbf{S}$ where $\mathbf{S}$ represents three (square) matrices which are the components of the intrinsic spin operator of the field (or particle). As a more familiar example, the Pauli spin operator [4] σ_x has the following effect on a two component wave function:

$$\sigma_x\psi = \begin{pmatrix} 0 & 1 \\ 1 & 0 \end{pmatrix}\begin{pmatrix} \psi_1 \\ \psi_2 \end{pmatrix} = \begin{pmatrix} \psi_2 \\ \psi_1 \end{pmatrix}$$

To emphasize the remark at the end of section 5, specification of ψ, the wave function, fixes the operators $\mathbf{S}$ within a unitary transformation. As will be fairly evident when particular cases are examined (section 21), we need to specify only the number of components of ψ.

Returning to our discussion of the orbital angular momentum operators, as indicated, we use the symbol $\mathbf{L}$ for them. Clearly,

$$\mathbf{L} = -i(\mathbf{r} \times \nabla) \tag{2.11}$$

Because the equation

$$(\nabla_i x_j - x_j \nabla_i)f = \delta_{ij}f, \qquad \nabla_i = \frac{\partial}{\partial x_i}$$

---

[4] In section 8 it will be seen that σ_x is twice the x-component of the angular momentum operator for an intrinsic spin $\frac{1}{2}$, etc. The form of σ_x given above corresponds to σ_z diagonal.

is true for all (differentiable) functions f, it may be written as an operator equation with the function f omitted:

$$\nabla_i x_j - x_j \nabla_i = \delta_{ij} \tag{2.12}$$

By using (2.11) and (2.12), the commutation relations for the orbital angular momenta can be obtained by straightforward algebra:

$$L_x L_y - L_y L_x = i L_z \qquad \text{(c.p.)} \tag{2.13}$$

Here and in the following c.p. stands for: and cyclic permutations; i.e.,

$$\mathbf{L} \times \mathbf{L} = i\,\mathbf{L}$$

The second line is simply a convenient summary of the first.

It will be shown in the next section that the definition of angular momentum operators (2.8)

$$R\psi = \psi' = e^{-i\theta(\mathbf{n}\cdot\mathbf{J})}\psi$$

in terms of the transformation properties of ψ leads to the same commutation relations for $\mathbf{J}$:

$$\mathbf{J} \times \mathbf{J} = i\,\mathbf{J} \tag{2.14}$$

The commutation rules and the specification that $\mathbf{J}$ be Hermitian provide an alternative definition of the angular momentum operators. However, we prefer the definition in terms of the generator of rotations of the coordinate axes, not only because this has a more immediate geometric meaning and is somewhat less formal, but also because in terms of this definition we can always construct an explicit realization of the $\mathbf{J}$-operators with a minimum of formalism, as soon as the field ψ is specified in a very general way. To be sure, the commutation rules also provide an opportunity for doing this, as section 8 shows, but the procedure seems somewhat less direct in that case.

7. COMMUTATION RULES FOR ANGULAR MOMENTUM OPERATORS

The angular momentum operators for a system was defined in section 5 in terms of the transformation properties of the wave function under rotations. A completely equivalent definition can be given in terms of the commutation relation $\mathbf{J} \times \mathbf{J} = i\,\mathbf{J}$. We shall now give a general proof that our definition of angular momentum implies these commutation relations. This has already been done for the special case of orbital angular momentum in section 6.

Let us apply two infinitesimal rotations to the coordinate system, $\exp(-i\,d\theta_x J_x)$ and $\exp(-i\,d\theta_y J_y)$, and investigate the difference that arises from applying them in different order. To determine the final orientation of the x-axis, for example, we fix our attention on a point P [with Cartesian coordinates $(1, 0, 0)$], that is, the point at a distance

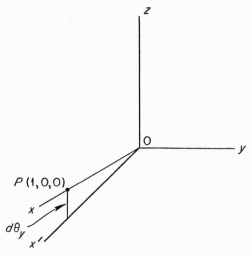

Fig. 1(a). Displacement of a point P as a result of rotations about the x- and then about the y-axis. The rotations are infinitesimal, and arcs of circles have been replaced by straight lines.

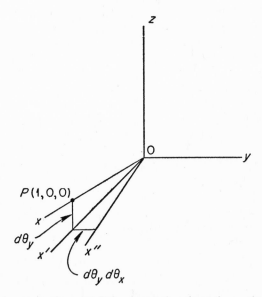

Fig. 1(b). Displacement of a point P due to rotation about the y-axis and then about the x-axis. Straight lines replace circular arcs as in Fig. 1(a).

unity from the origin, along the x-axis. Figure 1a shows the effect of first rotating through $d\theta_x$ about x and then rotating $d\theta_y$ about y: i.e., of the rotations $\exp\left(-i\,d\theta_y\,J_y\right)\exp\left(-i\,d\theta_x\,J_x\right)$. The first rotation, being about x, does not change the orientation of the x-axis. Figure 1b shows the effect of interchanging the order of the rotations. First we rotate $d\theta_y$ about y and then $d\theta_x$ about x. The rotation operator is $\exp\left(-i\,d\theta_x\,J_x\right)$ $\exp\left(-i\,d\theta_y\,J_y\right)$. Comparison of the two figures shows that the difference in the order of rotations is a displacement perpendicular to the original x-axis, the magnitude of which is a second-order infinitesimal. Hence the familiar statement that infinitesimal rotations of first order commute. For simplicity, in the above diagrams we have represented the displacements as straight lines rather than arcs of circles. The error is a second-order displacement, and it is immaterial in this discussion since the same error is made in both diagrams. The net displacement perpendicular to the x-axis is equivalent to a rotation about z of magnitude $d\theta_x\,d\theta_y$. Equating the net changes in the vector OP, we have

$$\exp\left(-i\,d\theta_x\,J_x\right)\exp\left(-i\,d\theta_y\,J_y\right) - \exp\left(-i\,d\theta_y\,J_y\right)\exp\left(-i\,d\theta_x\,J_x\right)$$
$$= \exp\left(-i\,d\theta_x\,d\theta_y\,J_z\right) - 1$$

We now expand the exponentials up to second-order in the infinitesimals $d\theta_x$ and $d\theta_y$.

$$[1 - i\,d\theta_x\,J_x - \tfrac{1}{2}(d\theta_x)^2\,J_x^2][1 - i\,d\theta_y\,J_y - \tfrac{1}{2}(d\theta_y)^2\,J_y^2]$$
$$- [1 - i\,d\theta_y\,J_y - \tfrac{1}{2}(d\theta_y)^2\,J_y^2][1 - i\,d\theta_x\,J_x - \tfrac{1}{2}(d\theta_x)^2\,J_x^2]$$
$$= -i\,d\theta_x\,d\theta_y\,J_z$$

The linear terms on the left cancel, and the second-order terms give

$$-d\theta_x\,d\theta_y\left(J_xJ_y - J_yJ_x\right) = -i\,d\theta_x\,d\theta_y\,J_z$$

Since this equation holds for arbitrary magnitudes of the rotations $d\theta_x$ and $d\theta_y$, we find the usual commutation rule for the angular momentum operators J_x and J_y:

$$J_xJ_y - J_yJ_x = iJ_z \tag{2.14a}$$

The other commutation relations follow from considering infinitesimal rotations about the other two pairs of coordinate axes.

8. EIGENVALUES OF THE ANGULAR MOMENTUM OPERATORS

It is the purpose of this section to show what quantities can be measured for an isolated system. Thus, the considerations of this section refer to a single system with total angular momentum j (in the usual sense of what an experimentalist refers to by this term). Part of our objective is to investigate precisely what a statement about the angular momentum implies.

The system can be an elementary particle (electron, proton, neutron) or an ensemble of such particles such as a nucleus. It can also be an atom or a group of atoms such as a molecule or crystal. As explained below, we can define (simultaneously) both a j and the value of any one component of the angular momentum with complete precision. In the absence of magnetic or electric fields, the states of the system corresponding to different angular momentum components are degenerate in energy, and all such states are equally populated. Under ordinary circumstances it is therefore useful to specify only the j of a state. As this discussion indicates, the degeneracy referred to is a consequence of the isotropy of space, and this will be evident from a physical point of view.

To establish the validity of these remarks, it is necessary to show that the operators corresponding to quantities simultaneously measurable with the energy commute with the Hamiltonian H, as well as with each other. In the physical problems with which we are concerned, the Hamiltonian is invariant under rotations. This is what we mean by an isolated (closed) system. Therefore, transformation of the Hamiltonian by the operator for a rotation about the x_k axis regenerates the Hamiltonian itself:

$$e^{-iJ_k\theta}\, H\, e^{iJ_k\theta} = H$$

or

$$e^{iJ_k\theta}\, H - H\, e^{iJ_k\theta} = 0$$

Expanding and setting every power of θ equal to zero shows that H commutes with any power of J_k,

$$H J_k^n - J_k^n H = 0 \tag{2.15}$$

Of course $HJ_k - J_k H = 0$ is sufficient to prove (2.15) for any n. In particular, H commutes with any component J_k and with the square $\mathbf{J}^2 = J_x^2 + J_y^2 + J_z^2$ of the angular momentum. Using square brackets for the commutator, that is, $[A, B] \equiv AB - BA$, we can write

$$[H, J_k] = 0$$
$$[H, \mathbf{J}^2] = 0 \tag{2.16a}$$

In addition, from (2.14) we see that

$$\begin{aligned}
[\mathbf{J}^2, J_x] &= [J_y^2 + J_z^2, J_x] \\
&= J_y[J_y, J_x] + [J_y, J_x]J_y \\
&\quad + J_z[J_z, J_x] + [J_z, J_x]J_z \\
&= -i(J_y J_z + J_z J_y) + i(J_z J_y + J_y J_z) = 0
\end{aligned}$$

By cyclic permutation it is clear that

$$[\mathbf{J}^2, J_k] = 0 \qquad (2.16b)$$

Thus, in view of (2.14), it is possible to diagonalize, along with H, the square and any one component of the angular momentum. We reiterate: Only one component can be diagonalized because the necessary and sufficient condition that a set of operators be diagonalized simultaneously is that all the operators commute; the one that is chosen diagonal is the z-component by definition.[5]

We now proceed to determine the eigenvalues. Let η_j and m be eigenvalues for $\mathbf{J}^2$ and J_z, respectively, and ψ_{jm} the eigenfunction; the energy and other eigenvalues are suppressed in the notation. Then,

$$\mathbf{J}^2 \, \psi_{jm} = \eta_j \, \psi_{jm}$$
$$J_z \, \psi_{jm} = m \, \psi_{jm} \qquad (2.17)$$

At this point j is merely an index to distinguish one eigenvalue of $\mathbf{J}^2$ from another. Now $J_x^2 + J_y^2 = \mathbf{J}^2 - J_z^2$ is diagonal with non-negative eigenvalues.

$$(J_x^2 + J_y^2) \, \psi_{jm} = (\eta_j - m^2) \, \psi_{jm}$$

where $\eta_j - m^2 \geq 0$. We have used the fact that the diagonal elements of the square of a Hermitian operator are positive or zero. This is seen as follows:

$$(\Omega^2)_{nn} = \sum_s \Omega_{ns}\Omega_{sn} = \sum_s \Omega_{sn}^*\Omega_{sn} = \sum_s |\Omega_{sn}|^2 \geq 0$$

This statement, of course, applies a fortiori to the sum of squares of Hermitian operators. Thus, for a given value of the total angular momentum and therefore a given η_j, the values of the projection on the z-axis, m, are bounded by $\pm\gamma_j$ where $0 \leq \gamma_j \leq \eta_j^{\frac{1}{2}}$.

We introduce the operators

$$J_\pm = J_x \pm i J_y \qquad (2.18)$$

with the commutation rules

$$[\mathbf{J}^2, J_\pm] = 0$$
$$[J_z, J_\pm] = \pm J_\pm \qquad (2.19)$$

---

[5] It should not be assumed that in all cases these are the only operators that can be diagonalized simultaneously. For example, if we describe a physical system such as an electron in a central field (in the non-relativistic approximation) we find that the total, orbital, and spin angular momenta are all diagonal (i.e., j^2, l^2, s^2 with $\mathbf{j} = \mathbf{l} + \mathbf{s}$, $\mathbf{l} = -i\mathbf{r} \times \nabla$, $\mathbf{s} = \frac{1}{2}\sigma$). The fact that we know this to be only an approximate description of an electron is irrelevant. In a more exact (relativistic) description j^2 and s^2 are diagonal along with j_z, and l^2 is not; see section 31.

as can be seen from (2.14) and (2.16b). We also introduce the functions

$$\phi_{jm}^{(\pm)} = J_\pm \psi_{jm} \tag{2.20}$$

These functions are eigenfunctions of $\mathbf{J}^2$ and J_z with eigenvalues η_j and $(m \pm 1)$, respectively. To see this we consider

$$\mathbf{J}^2(J_\pm \psi_{jm}) = \{J_\pm \mathbf{J}^2 + [\mathbf{J}^2, J_\pm]\} \psi_{jm} = \eta_j(J_\pm \psi_{jm})$$

and

$$J_z(J_\pm \psi_{jm}) = \{J_\pm J_z + [J_z, J_\pm]\} \psi_{jm} = J_\pm(J_z \pm 1) \psi_{jm}$$
$$= (m \pm 1)(J_\pm \psi_{jm})$$

Thus, we confirm the statement made above; that is,

$$\mathbf{J}^2 \phi_{jm}^{(\pm)} = \eta_j \phi_{jm}^{(\pm)}$$
$$J_z \phi_{jm}^{(\pm)} = (m \pm 1) \phi_{jm}^{(\pm)} \tag{2.21}$$

and $\phi_{jm}^{(\pm)}$ must be proportional to the normalized eigenfunctions $\psi_{jm\pm1}$. We write

$$J_\pm \psi_{jm} = \phi_{jm}^{(\pm)} = \Gamma_\pm \psi_{jm\pm1} \tag{2.22}$$

Of course, the proportionality constant $\Gamma_\pm$ can be zero in some cases.

Because of (2.22) the operators $J_\pm$ are sometimes called the "raising" and "lowering" operators. Since the values of m are bounded for given j, there must be a largest value m_2 and a smallest m_1, such that

$$J_+ \psi_{jm_2} = 0$$
$$J_- \psi_{jm_1} = 0 \tag{2.22'}$$

Otherwise there would be a value of m greater than $\eta_j^{\frac{1}{2}}$ and a value less than $-\eta_j^{\frac{1}{2}}$ contrary to the conclusion already established. Applying J_- to the first of these equations and J_+ to the second gives

$$\eta_j - m_2(m_2 + 1) = 0$$
$$\eta_j - m_1(m_1 - 1) = 0 \tag{2.23}$$

where we have used

$$J_\mp J_\pm = \mathbf{J}^2 - J_z(J_z \pm 1) \tag{2.24}$$

Elimination of η_j in (2.23) yields

$$m_2(m_2 + 1) = m_1(m_1 - 1)$$

or

$$(m_2 + m_1)(m_2 - m_1 + 1) = 0$$

In view of the fact that $m_2 \geq m_1$, the only possible solution is

$$m_1 = -m_2$$

Since successive values of m differ by unity, $m_2 - m_1$ is a non-negative integer. We denote this integer by $2j$ so that j can have any of the following values:

$$j = 0, \tfrac{1}{2}, 1, \tfrac{3}{2}, 2, \cdots$$

That is, j can be an integer or a half-integer. Then, from $m_2 - m_1 = 2j$ and $m_1 = -m_2$ we find that

$$m_2 = j \qquad \text{and} \qquad m_1 = -j$$

For a given j the permitted values of m are $-j \le m \le j$; i.e.,

$$m = j, j-1, j-2, \cdots, -j+1, -j \qquad (2.24a)$$

Thus, there are $2j+1$ permissible values of m for each j. Inserting these results into either of the equations in (2.23) gives

$$\eta_j = j(j + 1) \qquad (2.24b)$$

Summarizing, we have

$$\mathbf{J}^2 \, \psi_{jm} = j(j + 1) \, \psi_{jm} \qquad j - \text{half or whole integral} \qquad (2.25)$$

$$J_z \, \psi_{jm} = m \, \psi_{jm} \qquad\qquad m = j, j-1, \cdots, -j$$

The number j is the angular momentum in units $\hbar$. Classically, $j \to \infty$, and then the eigenvalue of $\mathbf{J}^2$ is j^2. We also recognize that the maximum value of the J_z eigenvalue is j; so again the designation of j as "angular momentum" is reasonable. The extra term in the exact eigenvalue $j^2(1 + 1/j)$ is a quantum-mechanical effect arising from the non-commutability of the $\mathbf{J}$-operators; see (2.24). It is therefore seen to arise from the impossibility of precisely defining the *direction* of the angular momentum vector.

These results also show that, in a real sense, it might be more appropriate to talk about the value of the operator $\mathbf{J}^2$ to be assigned to the states of our isolated system. Since this depends on the parameter j, in a simple way and has a well known classical limit, it is just as convenient to refer to the value of j and to utilize various data to ascertain this number. Further discussion of this question is given in section 9.

Having found the eigenvalues of $\mathbf{J}^2$ and J_z, it is easy to find the matrix elements of J_x and J_y, or their equivalents $J_\pm$, in this representation in which $\mathbf{J}^2$ and J_z are diagonal. From (2.22) we have that

$$(J_\pm \, \psi_{jm}, J_\pm \, \psi_{jm}) = (\Gamma_\pm \, \psi_{j\, m\pm1}, \Gamma_\pm \, \psi_{j\, m\pm1}) = |\Gamma_\pm|^2 \, (\psi_{j\, m\pm1}, \psi_{j\, m\pm1})$$

and, since the ψ_{jm} are orthonormal,

$$|\Gamma_\pm|^2 = (J_\pm \, \psi_{jm}, J_\pm \, \psi_{jm})$$

The operators $J_\pm$ are not Hermitian although J_x and J_y are. Instead

$$(J_\pm)^+ = J_\mp \qquad (2.26)$$

Using (2.24), we get

$$|\Gamma_\pm|^2 = (\psi_{jm}, J_\mp J_\pm \psi_{jm}) = \{\psi_{jm}, [\mathbf{J}^2 - J_z(J_z \pm 1)] \psi_{jm}\}$$

or

$$|\Gamma_\pm|^2 = j(j+1) - m(m \pm 1) = (j \mp m)(j \pm m + 1)$$

Only the absolute value is determined, and the phase is arbitrarily chosen so that

$$\Gamma_\pm = [(j \mp m)(j \pm m + 1)]^{\frac{1}{2}} \qquad (2.27)$$

We are now able to write down all the matrix elements of the angular momentum operators in a representation in which $\mathbf{J}^2$ and J_z are diagonal. The matrix elements will be written in the form

$$(jm| \quad |j'm')$$

with the operator placed in the middle. Since all the angular momentum operators J_k are diagonal in j, see (2.21), matrices for each j are usually presented with the rows labeled by m and the columns by m'. The matrix elements are

$$(jm| \mathbf{J}^2 |j'm') = j(j+1)\, \delta_{jj'}\, \delta_{mm'}$$

$$(jm| J_z |j'm') = m\, \delta_{jj'}\, \delta_{mm'} \qquad (2.28)$$

$$(jm| J_\pm |j'm') = [(j \mp m')(j \pm m' + 1)]^{\frac{1}{2}}\, \delta_{jj'}\, \delta_{m,m' \pm 1}$$

In writing explicit matrices the convention is followed of placing the element for $m = j$ in the first row and $m' = j$ in the first column; thus,

$$
(jm|\ |j\,m'):\
\begin{array}{c}
\\
m = j \\
m = j - 1 \\
m = j - 2 \\
\vdots
\end{array}
\begin{array}{cc}
m' = j & m' = j - 1 \quad \cdots \\
\left[\begin{array}{ccc}
(j\,j|\ |j\,j) & (j\,j|\ |j\,j{-}1) & \cdots \\
(j\,j{-}1|\ |j\,j) & (j\,j{-}1|\ |j\,j{-}1) & \cdots \\
(j\,j{-}2|\ |j\,j) & & \\
\vdots & \vdots &
\end{array}\right]
\end{array}
$$

The matrix representation of the operators for $j = \frac{1}{2}$ and $j = 1$ are given below. $j = \frac{1}{2}$:

$$(\tfrac{1}{2}m| \mathbf{J}^2 |\tfrac{1}{2}m'): \quad \begin{pmatrix} \frac{3}{4} & 0 \\ 0 & \frac{3}{4} \end{pmatrix} = \tfrac{1}{2} \cdot (\tfrac{1}{2} + 1)\, I$$

I is the unit matrix with the appropriate dimensionality (here 2-by-2).

$$(\tfrac{1}{2}m|J_z|\tfrac{1}{2}m'):\quad \begin{pmatrix} \tfrac{1}{2} & 0 \\ 0 & -\tfrac{1}{2} \end{pmatrix} = \tfrac{1}{2}\sigma_z$$

$$(\tfrac{1}{2}m|J_+|\tfrac{1}{2}m'):\quad \begin{pmatrix} 0 & 1 \\ 0 & 0 \end{pmatrix}$$

$$(\tfrac{1}{2}m|J_-|\tfrac{1}{2}m'):\quad \begin{pmatrix} 0 & 0 \\ 1 & 0 \end{pmatrix}$$

The matrices for J_x and J_y obtained with the relations $J_x = \tfrac{1}{2}(J_+ + J_-)$ and $J_y = \dfrac{1}{2i}(J_+ - J_-)$ are

$$(\tfrac{1}{2}m|J_x|\tfrac{1}{2}m'):\quad \begin{pmatrix} 0 & \tfrac{1}{2} \\ \tfrac{1}{2} & 0 \end{pmatrix} = \tfrac{1}{2}\sigma_x$$

$$(\tfrac{1}{2}m|J_y|\tfrac{1}{2}m'):\quad \begin{pmatrix} 0 & -\tfrac{1}{2}i \\ \tfrac{1}{2}i & 0 \end{pmatrix} = \tfrac{1}{2}\sigma_y$$

The σ_k are the familiar Pauli spin matrices: They are written below for purposes of reference.

$$\sigma_x = \begin{pmatrix} 0 & 1 \\ 1 & 0 \end{pmatrix} \qquad \sigma_y = \begin{pmatrix} 0 & -i \\ i & 0 \end{pmatrix} \qquad \sigma_z = \begin{pmatrix} 1 & 0 \\ 0 & -1 \end{pmatrix}$$

Special commutations rules apply to these: namely,

$$\sigma_j \sigma_k = i\,\sigma_l$$

with j, k, and l a cyclic permutation of x, y, z.

$$j = 1: \qquad (1m|\mathbf{J}^2|1m'):\quad \begin{bmatrix} 2 & 0 & 0 \\ 0 & 2 & 0 \\ 0 & 0 & 2 \end{bmatrix} = 1(1+1)\,I$$

$$(1m|J_z|1m'):\quad \begin{bmatrix} 1 & 0 & 0 \\ 0 & 0 & 0 \\ 0 & 0 & -1 \end{bmatrix}$$

$$(1m|J_+|1m'):\quad \begin{bmatrix} 0 & \sqrt{2} & 0 \\ 0 & 0 & \sqrt{2} \\ 0 & 0 & 0 \end{bmatrix}$$

$$(1m|J_-|1m'):\quad \begin{bmatrix} 0 & 0 & 0 \\ \sqrt{2} & 0 & 0 \\ 0 & \sqrt{2} & 0 \end{bmatrix}$$

From these we get

$$(1m|J_x|1m'): \quad \frac{1}{\sqrt{2}} \begin{bmatrix} 0 & 1 & 0 \\ 1 & 0 & 1 \\ 0 & 1 & 0 \end{bmatrix}$$

and

$$(1m|J_y|1m'): \quad \frac{i}{\sqrt{2}} \begin{bmatrix} 0 & +1 & 0 \\ -1 & 0 & +1 \\ 0 & -1 & 0 \end{bmatrix}$$

9. PHYSICAL INTERPRETATION OF ANGULAR MOMENTUM

At this stage of the development of the theory it is pertinent to inquire in somewhat greater detail into the relationship between the eigenvalues of the operators we have introduced under the name angular momentum and the corresponding quantities which are measured in various kinds of experiments (for example, an atomic beam experiment, a measurement of hyperfine structure, or angular correlation measurement) and which are customarily referred to by that name.

We are primarily concerned with the angular momentum of quantum systems. The description of classical systems would then follow from the Correspondence Principle. This preoccupation with quantum systems is understandable in view of current developments in physics. But it is important to realize that this concern commits us to a more general point of view in formulating a concept of angular momentum than would otherwise be the case. This arises from the circumstance that only the orbital part of the angular momentum has a classical counterpart. On the other hand, it is desirable and eventually necessary to define angular momentum in a general way so that both orbital and intrinsic parts appear on an equal footing. In this way a unified point of view is possible, and at the same time a better understanding of the distinction between them can be reached.

Despite these remarks we begin with a discussion of the orbital angular momentum because the connection with the familiar domain of classical physics is comparatively direct. We start with the classical definition of the vector angular momentum, $\mathbf{r} \times \mathbf{p}$, where $\mathbf{p}$ is the linear momentum. Following the usual description we define a set of three operators, L_x, L_y, and L_z, or in short $\mathbf{L}$, by replacing $\mathbf{p}$ by $-i\nabla$ (recalling that we set $\hbar = 1$). An immediate difference relative to the classical situation is the failure of the commutators to vanish; i.e.,

$$\mathbf{L} \times \mathbf{L} = i\,\mathbf{L}$$

as one finds directly by use of $[\nabla_j, x_k] = \delta_{jk}$. The customary interpretation of this result, that different components of the (orbital) angular momentum cannot be simultaneously measured with precision, means that quantum states cannot include among their specifying labels the values of any two components. A meaningful set of labels (quantum numbers) is obtained from the constants of the motion: the eigenvalues of operators which commute with the Hamiltonian and with each other. For systems which can be described in terms of rotational isotropy (particles with no intrinsic spin in central fields), $L^2 = L_x^2 + L_y^2 + L_z^2$ and one component, L_z say, have this property. Hence, a state includes among other labels the numbers l and m, corresponding to the eigenvalues of these operators. It will be noted that here, as in the classical case, the time-independence of the orbital angular momentum l is a direct consequence of the invariance of the Hamiltonian under rotations. This means that for the system under considerations all directions in space are equivalent and space is isotropic.

In a purely formal way we can state that all the properties of a system (including the isotropy of space) are contained in the wave function ψ_{lm} which describes it. In fact, if ψ_{lm} were given, the particular value of l would be found by applying $L^2 \psi_{lm} = l(l + 1) \psi_{lm}$. It should be understood that in performing this operation we are investigating the properties of ψ_{lm}, and the physical system which this function describes, under rotations of the coordinate axes. In fact, we could have defined the components of $\mathbf{L}$ in terms of the effect on ψ_{lm} of the rotations discussed in section 6. This procedure, though it seems less direct than the one followed in this section, is the one that must be the basis of a general treatment of angular momentum. Thus, we are free of the limitations connected with the lack of a classical counterpart of intrinsic spin if angular momentum is defined generally in terms of the effect on the wave functions of rotations of the coordinate system. If this is done for fields that depend only on the spatial coordinates we are restricted to the orbital operators. If additional degrees of freedom (finite in number) are specified, the intrinsic spin appears. It is important to remember that these degrees of freedom are associated with rotations in ordinary space. As soon as the nature of the variables describing the system is specified, the structure of the orbital angular momentum operators is defined. Examples have already been given, and further illustrations will appear in the sequel.

The physical interpretation of angular momentum rests on the possibility of making a connection between the results of a measurement and the eigenvalues of the angular momentum operators. The latter, as we have seen, are connected with the rotational symmetry properties of a physical system as expressed in the wave function describing it. In almost every measurement, however, the quantity observed seems to have no *direct* connection with the rotational properties. We mention two kinds of measurements: total intensity and angular distribution or correlation data.

The first is exemplified by total cross section measurements which can yield a value of the total angular momentum j, of a nuclear energy state say, because the cross section depends on the number of ways $(2j + 1)$ in which this state can be formed. This results from the fact that under usual circumstances (absence of magnetic fields or other direction-indicating perturbations) the $2j+1$ substates corresponding to each possible orientation of the angular momentum are all degenerate. The fact that this enters the cross section is, of course, a result of the theory, and the assumption of any particular value of j (such as must be made in comparison of the calculated cross section with the data) is equivalent to a specific assumption as to rotational properties of the state involved.[6]

This point is made even more clearly if we examine the way in which angular momenta are determined from angular correlation data. The angular correlation process will be discussed in section 33 of part B. However, it will suffice to note here that the coincidence counting rate of two detectors as a function of angle between the two radiations observed depends on the angular momenta of the states involved for one reason only: At a certain point in the theory it is necessary to specify the rotational transformation properties of these states. In fact, so far as the emitting and/or absorbing states are concerned, no other property need be specified.

To summarize, a measurement of angular momentum is always made by comparing the results of a measurement with the results predicted by theory, and in the latter the angular momentum parameters occur because a statement has been made that the states involved have perfectly definite properties under rotations of the coordinate axes. Therefore, the significance of all such measurements is to be found in terms of statements concerning rotational symmetry.[7]

[6] The same is true in determining j from counting hyperfine structure components. Here the number of different resultants that can be formed from adding two angular momenta is involved; see section 10 below.

[7] In this connection it is of interest to call attention to one particular experiment designed to measure the angular momentum of light. Here one measures the torque communicated to a doubly refracting plate which changes the state of polarization of the (circularly or elliptically polarized) light beam. This experiment is described by R. A. Beth, *Phys. Rev.* **48**, 471 (1935); **50**, 115 (1936). In this experiment only the component of angular momentum along the propagation direction is involved. This would then be a measurement of the intrinsic spin of a photon (section 21). The effect is a classical one. Moreover, the necessity of an interpretation in terms of rotational properties is not apparent. However, from an operational point of view this does not detract from the validity of any statements made above.

III. COUPLING OF TWO ANGULAR MOMENTA

10. DEFINITION OF THE CLEBSCH-GORDAN COEFFICIENTS

Once we have obtained the eigenfunctions which diagonalize the square and z-component of the angular momentum, there arises the problem of finding the wave functions for a compound system of *two* angular momenta. The necessity for compounding angular momenta arises when we deal with a single particle whose total angular momentum is a sum of two parts (orbital and intrinsic spin), or with two or more particles (see also Chapters X and XI) and also when we consider scattering or emission/absorption processes between states of well-defined angular momentum.

Let the eigenfunctions for the angular momenta j_1 and j_2 be $\psi_{j_1 m_1}$ and $\psi_{j_2 m_2}$, respectively. Then

$$\mathbf{J}_1^2 \, \psi_{j_1 m_1} = j_1(j_1 + 1) \, \psi_{j_1 m_1}; \qquad \mathbf{J}_2^2 \, \psi_{j_2 m_2} = j_2(j_2 + 1) \, \psi_{j_2 m_2} \tag{3.1}$$
$$J_{1z} \, \psi_{j_1 m_1} = m_1 \, \psi_{j_1 m_1}; \qquad J_{2z} \, \psi_{j_2 m_2} = m_2 \, \psi_{j_2 m_2}$$

To keep the notation simple, the same symbol ψ has been used for both eigenfunctions, even though they refer to completely different spaces and may have different structure as in the case of orbital and intrinsic spin eigenfunctions. We shall call the direct product

$$\psi_{j_1 m_1} \, \psi_{j_2 m_2} \tag{3.2}$$

the uncoupled representation. In the uncoupled representation $\mathbf{J}_1^2$, J_{1z} and $\mathbf{J}_2^2$, J_{2z} are diagonal.[1] The total angular momentum (vector) operator $\mathbf{J}$ is defined by

$$\mathbf{J} = \mathbf{J}_1 + \mathbf{J}_2 \tag{3.3}$$

That the sum of two angular momenta is also an angular momentum

[1] In section 3 we mentioned that this might be the case for a strong external magnetic field decoupling the two angular momenta, so that the two spins are unaware of each other's presence. If there is no magnetic field and there is dipole coupling ($\mathbf{J}_1 \cdot \mathbf{J}_2$) of the two systems, the diagonal operators are $\mathbf{J}_1^2$, $\mathbf{J}_2^2$, $\mathbf{J}^2$, and J_z, where $\mathbf{J} = \mathbf{J}_1 + \mathbf{J}_2$. Thus, J_z is diagonal in both representations since both situations correspond to symmetry around the z-axis.

may be proved by showing that the commutation relation $\mathbf{J} \times \mathbf{J} = i\mathbf{J}$ is satisfied. Thus,

$$[J_x, J_y] = [J_{1x}+J_{2x}, J_{1y}+J_{2y}]$$

$$= [J_{1x}, J_{1y}] + [J_{1x}, J_{2y}] + [J_{2x}, J_{1y}] + [J_{2x}, J_{2y}]$$

Since angular momenta in different spaces commute, the two middle brackets vanish and

$$[J_x, J_y] = iJ_{1z} + iJ_{2z} = iJ_z \quad \text{(c.p.)}$$

The commutation rules are therefore valid for the sum of any number of angular momenta.

We seek a representation in which $\mathbf{J}^2$ and J_z, as well as $\mathbf{J}_1^2$ and $\mathbf{J}_2^2$, are diagonal with eigenvalues $j(j+1)$ and m, respectively. This coupled representation ψ_{jm} is connected with the uncoupled representation of (3.2) by a unitary transformation

$$\psi_{jm} = \sum_{m_1 m_2} C(j_1 j_2 j; m_1 m_2 m) \, \psi_{j_1 m_1} \psi_{j_2 m_2} \qquad (3.4)$$

The elements of the transformation $C(j_1 j_2 j; m_1 m_2 m)$ are called Clebsch–Gordan (or Wigner) coefficients. For brevity we refer to them as "C-coefficients." They are also often referred to as vector-addition coefficients, since we add j_1 to j_2 to get j, and m_1 to m_2 to get m.[2]

If we apply $J_z = J_{1z} + J_{2z}$ to (3.4), we get

$$m \, \psi_{jm} = \sum_{m_1 m_2} (m_1 + m_2) \, C(j_1 j_2 j; m_1 m_2 m) \, \psi_{j_1 m_1} \psi_{j_2 m_2}.$$

or

$$\sum_{m_1 m_2} (m - m_1 - m_2) \, C(j_1 j_2 j; m_1 m_2 m) \, \psi_{j_1 m_1} \psi_{j_2 m_2} = 0$$

Since the $\psi_{j_1 m_1} \psi_{j_2 m_2}$ are linearly independent, this sum cannot vanish unless the coefficient of each term is identically zero:

$$(m - m_1 - m_2) \, C(j_1 j_2 j; m_1 m_2 m) = 0$$

---

[2] As we shall see presently, the addition of j_1 and j_2 is vectorial and that of m_1 and m_2 is algebraic. The question will arise as to the distinction between addends and resultant. That is, in what respect would the situation differ if j_1 and j were added to obtain j_2. This question is answered by the symmetry relations of the Clebsch–Gordan coefficients given in section 11. The definition of the C-coefficients also implies an order of the addends, but it will be seen that interchanging j_1 and j_2 implies only a phase change. That a phase should be expected is clear. We also recognize that, in going from

$$\mathbf{J}_1 + \mathbf{J}_2 = \mathbf{J} \qquad \text{to} \qquad \mathbf{J}_1 + \mathbf{J} = \mathbf{J}_2$$

we are changing the sign of $\mathbf{J}$ and $\mathbf{J}_2$ or of $\mathbf{J}_1$. The commutation rules (2.14) have $-i$ in place of i when the signs of all components of an angular momentum operator are changed.

Thus

$$C(j_1 j_1 j; m_1 m_2 m) = 0 \quad \text{unless} \quad m = m_1 + m_2 \qquad (3.5)$$

Therefore, not all three projection quantum numbers are independent, and the double sum in (3.4) is actually a single sum. We shall usually take advantage of this by replacing m_2 by $m - m_1$ and suppressing the third projection quantum number in the C-coefficient:

$$\psi_{jm} = \sum_{m_1} C(j_1 j_2 j; m_1, m-m_1)\, \psi_{j_1 m_1} \psi_{j_2\, m-m_1} \qquad (3.6)$$

It should be remembered, however, that the third projection quantum number is always the sum of the first two.

The C-coefficients are elements of a unitary transformation and thus must satisfy orthogonality relations of the type given in equations (1.12) and (1.13). These can be simply derived from the fact that the eigenfunctions in (3.6) are all orthonormal:

$$(\psi_{jm}, \psi_{j'm'}) = \delta_{jj'}\, \delta_{mm'}$$

The standard phase convention made for the C-coefficients is such that they are real. In that case

$$\sum_{m_1} \sum_{m_1'} C(j_1 j_2 j; m_1, m-m_1)\, C(j_1 j_2 j'; m_1', m'-m_1')(\psi_{j_1 m_1}, \psi_{j_1 m_1'})$$

$$\times\ (\psi_{j_2\, m-m_1}, \psi_{j_2\, m'-m_1'}) = \delta_{jj'}\, \delta_{mm'}$$

The two values j and j' are both obtained from the same values of j_1 and j_2. The $\psi_{j_1 m_1}$ and $\psi_{j_2 m_2}$ are also orthonormal. Thus,

$$(\psi_{j_1 m_1}, \psi_{j_1 m_1'})(\psi_{j_2\, m-m_1}, \psi_{j_2\, m'-m_1'}) = \delta_{m_1 m_1'}\, \delta_{mm'}$$

Therefore

$$\sum_{m_1} C(j_1 j_2 j; m_1, m-m_1)\, C(j_1 j_2 j'; m_1, m-m_1) = \delta_{jj'} \qquad (3.7)$$

With this orthogonality relation [3] we can show that the inverse expansion of (3.4) is

$$\psi_{j_1 m_1} \psi_{j_2\, m-m_1} = \sum_{j} C(j_1 j_2 j; m_1, m-m_1)\, \psi_{jm} \qquad (3.8)$$

---

[3] If the elements of a unitary matrix are real, the matrix is orthogonal. Note the structure of (3.7). In it both the first and second projection numbers vary but in such a way that their sum is constant. The first pair of these numbers must match in the two C-coefficients. This is also true for the angular momenta which are written as j_1 and j_2 in (3.7). Compare with the structure of (3.9) below.

Both sides of (3.8) are multiplied by $C(j_1 j_2 j'; m_1, m - m_1)$ and summed over m_1,

$$\sum_j \psi_{jm} \left(\sum_{m_1} C(j_1 j_2 j; m_1, m - m_1) \, C(j_1 j_2 j'; m_1, m - m_1) \right)$$
$$= \sum_{m_1} C(j_1 j_2 j'; m_1, m - m_1) \, \psi_{j_1 m_1} \psi_{j_2 \, m - m_1}$$

The quantity in parentheses is, according to (3.7), just $\delta_{jj'}$; so

$$\psi_{j'm} = \sum_{m_1} C(j_1 j_2 j'; m_1, m - m_1) \, \psi_{j_1 m_1} \, \psi_{j_2 \, m - m_1}$$

which is the same as (3.6). From (3.8) another orthogonality condition can be deduced in the same way as (3.7) was obtained. This will be seen to be an expression of the orthonormality condition on the rows, as well as columns, of a unitary matrix.

$$\sum_j C(j_1 j_2 j; m_1, m - m_1) \, C(j_1 j_2 j; m_1', m' - m_1') = \delta_{m_1 m_1'} \delta_{mm'} \quad (3.9)$$

Nothing has yet been said about the range of j and m. Quite generally we know that, for given j,

$$-j \leq m \leq j \qquad \text{and} \qquad m_{\max} = j$$

Since $m = m_1 + m_2$ the maximum value of m for all j is $j_1 + j_2$; i.e., the maximum value of $m_1 + m_2$. This must be the maximum value of j also, for otherwise there would be a larger value of $m_1 + m_2$. Thus,

$$j_{\max} = j_1 + j_2 \qquad (3.10)$$

In the "stretched" case, where j and m have this maximum value, the relation between the coupled representation and the uncoupled representation is especially simple because there is only one value of m_1 (and of m_2) permissible. Equation (3.6) reduces to one term which can be fixed within a phase by the normalization requirement.

$$\psi_{j_1 + j_2 \, j_1 + j_2} = \epsilon \, \psi_{j_1 j_1} \psi_{j_2 j_2}$$
$$C(j_1 j_2 \, j_1 + j_2; j_1 j_2) = \epsilon, \qquad |\epsilon|^2 = 1 \qquad (3.11)$$

We follow the standard convention: $\epsilon = 1$; i.e.,

$$C(j_1 j_2 \, j_1 + j_2; j_1 j_2) = 1 \qquad (3.11a)$$

The next smaller value of m is $j_1 + j_2 - 1$ arising from the two uncoupled states $\psi_{j_1 \, j_1 - 1} \psi_{j_2 j_2}$ and $\psi_{j_1 j_1} \psi_{j_2 \, j_2 - 1}$. One combination of these gives the coupled state $\psi_{j_1 + j_2 \, j_1 + j_2 - 1}$ while the other can give only the coupled state $\psi_{j_1 + j_2 - 1 \, j_1 + j_2 - 1}$ with $m = j = j_1 + j_2 - 1$. Similarly, from the three uncoupled states for $m = j_1 + j_2 - 2$, $\psi_{j_1 j_1} \psi_{j_2 j_2 - 2}$, $\psi_{j_1 j_1 - 1} \psi_{j_2 j_2 - 1}$, and $\psi_{j_1 j_1 - 2} \psi_{j_2 j_2}$ are obtained the three coupled states $\psi_{j_1 + j_2 \ j_1 + j_2 - 2}$, $\psi_{j_1 + j_2 - 1 \ j_1 + j_2 - 2}$, and $\psi_{j_1 + j_2 - 2 \ j_1 + j_2 - 2}$; for the last state $m = j = j_1 + j_2 - 2$. Each time m is reduced by unity in this counting procedure, one of the coupled states which arises belongs to a j-value reduced by unity, and in this state m has its maximum value. The lower limit $j_{\min}$ is reached when the number of coupled substates has been matched against all of the uncoupled states. Then

$$\sum_{j_{\min}}^{j_{\max}} (2j + 1) = (2j_1 + 1)(2j_2 + 1) \tag{3.12}$$

This counting procedure will determine $j_{\min}$ since $j_{\max} = j_1 + j_2$. The left side of (3.12) can be evaluated using

$$\sum_{\alpha}^{\beta} j = \tfrac{1}{2}[\beta(\beta + 1) - \alpha(\alpha - 1)]$$

whether j, as well as α and β, is an integer or a half-integer. Then we find

$$j_{\min}^2 = (j_1 - j_2)^2$$

and, since $j_{\min} \geq 0$,

$$j_{\min} = |j_1 - j_2| \tag{3.13}$$

Therefore, if $\mathbf{J} = \mathbf{J}_1 + \mathbf{J}_2$, and if the eigenvalues of $\mathbf{J}_1^2$ are $j_1(j_1 + 1)$ and the eigenvalues of $\mathbf{J}_2^2$ are $j_2(j_2 + 1)$, then the eigenvalues of $\mathbf{J}^2$ are $j(j + 1)$ where

$$j = j_1 + j_2, j_1 + j_2 - 1, \cdots |j_1 - j_2|$$

The numbers j_1, j_2, and j are said to form a triangle, and this relation is denoted by $\Delta(j_1 j_2 j)$, which is symmetric in the three angular momenta. The Clebsch–Gordan coefficients must vanish if this condition is not fulfilled, and the results obtained below verify that, in agreement with the definition, this is indeed the case. Thus it is not necessary to specify the limits of the sums like those in (3.7) or (3.9), for the terms outside the permissible limits will contain vanishing C-coefficients.

$$C(j_1 j_2 j; m_1, m - m_1) = 0 \qquad \text{unless} \qquad \Delta(j_1 j_2 j) \tag{3.14}$$

In addition, we have the restrictions

$$|m_1| \leq j_1, \qquad |m| \leq j, \qquad |m - m_1| \leq j_2 \tag{3.15}$$

The sums over projection quantum numbers contain $2j_s + 1$ terms, where j_s is the smaller of j_1 and j_2. This can be seen from (3.6) where we could have eliminated m_1 and summed over m_2 instead of eliminating m_2 and summing over m_1.

We conclude this section with a few words on notation. There are an almost embarrassing number of notations for the C-coefficients in the literature. Condon and Shortley [4] use $(j_1 j_2 m_1 m_2 | j_1 j_2 jm) = (j_1 j_2 jm | j_1 j_2 m_1 m_2)$ for our $C(j_1 j_2 j; m_1, m - m_1)$, $(m_2 = m - m_1)$. This notation has the advantage of clearly indicating what is diagonal in the two representations corresponding to the two coupling schemes. However, it is slightly more cumbersome and, from an information viewpoint, rather redundant. Again, in favor of the Condon–Shortley notation is the fact that it is a natural way to write a scalar product of ψ_{jm} and $\psi_{j_1 m_1} \psi_{j_2 m_2}$—if ψ_{jm}^* is denoted by $(jm|$ and $\psi_{j_1 m_1}$ and $\psi_{j_2 m_2}$ by $|j_1 m_1)$ and $|j_2 m_2)$, respectively. However, the logical extension of this procedure would be the appearance (in the equations to follow) of a wide variety of brackets of various kinds. Presumably these would be distinguished by round, square, curly, and angular brackets along with single, double, etc. vertical bars which will also be introduced in matrix elements. At the same time these brackets would have to be distinguished from brackets signifying multiplication, and, as a consequence, confusion is sometimes difficult to avoid. Other notations have appeared in the literature. Those with superscripts and subscripts, to both of which subscripts are attached, are avoided for technical reasons of reproduction. Still another notation $(m_1 m_2 | jm)_{j_1 j_2}$ is sufficiently compact but does not emphasize the degree of symmetry which exists among the three angular momenta constituting the triangle. This is more than a matter of taste because use of the symmetry relations given in section 11 is facilitated by a more symmetric notation. The same remark applies to the notation $(m_1 m_2 | jm)$.[4] Our notation has a redundant symbol C. We have retained this symbol under the impression that this makes the equations easier to read. A comma will be inserted between the two projection quantum numbers in the C-coefficient whenever ambiguity might otherwise result.

11. SYMMETRY RELATIONS OF THE CLEBSCH–GORDAN COEFFICIENTS

We consider the addition of angular momenta as in section 10 but with j, m replaced by j_3, m_3 to emphasize the symmetry properties. It is evident that among the three number pairs j_1, m_1; j_2, m_2; and j_3, $m_3 = m_1 + m_2$, there is a certain symmetry as evidenced by the triangular relation between j_1, j_2, and j_3 and the m-sum rule. We may therefore expect that some simple relations exist between C-coefficients when the

[4] E. U. Condon and G. H. Shortley, *Theory of Atomic Spectra*, Cambridge University Press, 1935.

roles of the participating angular momenta are interchanged.[2] As defined, the C-coefficients possess a higher degree of symmetry between j_1 and j_2 than between j_3 and either of the other angular momenta. For example, a change in the order in which j_1 and j_2 are compounded to give the resultant j will introduce only a change in phase in the C-coefficient. On the other hand, interchange of j_1 or j_2 with j_3 introduces factors depending on statistical weights, $2j_1 + 1$, etc.; see (3.17) below. This asymmetry is not essential, and a redefinition of the coefficients, see (3.21), would remove this unbalance. For historical reasons and, more important, to make comparison with the literature easier, we place the main emphasis on the C-coefficients as defined above. It must be understood that, whenever the total angular momenta are interchanged, their projections must also be interchanged but with such sign changes as are needed to keep the sum rule $m_1 + m_2 = m_3$ inviolate. The symmetry relations have been derived independently by Racah [5] and Eisenbud [6] from explicit expressions for the C-coefficients. Three independent symmetry relations are

$$C(j_1 j_2 j_3; m_1 m_2 m_3) = (-)^{j_1+j_2-j_3} C(j_1 j_2 j_3; -m_1, -m_2, -m_3) \quad (3.16a)$$

$$= (-)^{j_1+j_2-j_3} C(j_2 j_1 j_3; m_2 m_1 m_3) \quad (3.16b)$$

$$= (-)^{j_1-m_1} \left(\frac{2j_3 + 1}{2j_2 + 1}\right)^{\frac{1}{2}} C(j_1 j_3 j_2; m_1, -m_3, -m_2) \quad (3.16c)$$

These results are derived below from an explicit expression for the C-coefficients, and they will be used to derive all further results.

The phases in the relations (3.16) are all real since by virtue of $\Delta(j_1 j_2 j_3)$, $j_1+j_2+j_3$ is an integer; moreover, $(-)^n = (-)^{-n}$ for integral n. Thus $j_1+j_2-j_3$ is also an integer and $(-)^{j_1+j_2-j_3}$ is real. The first relation (3.16a) shows that an overall change in the sign of the projection quantum numbers is equivalent to a change in phase. According to (3.16b), the same change of phase is obtained on interchanging the roles of the two angular momenta j_1 and j_2 which are being added, together with their projections m_1 and m_2. The third symmetry relation (3.16c) shows that, when the resultant j is interchanged with an addend, here j_2, the square root of a statistical weight ratio as well as a phase factor is introduced. The lack of symmetry is not surprising in view of the special role played by the resultant j_3 and its projection m_3; however,

[5] G. Racah, *Phys. Rev.* **62**, 438 (1942).

[6] L. Eisenbud, Ph.D. Thesis, Princeton University, 1948. Also E. P. Wigner, On the Matrices Which Reduce the Kronecker Products of Representations of Simply Reducible Groups (unpublished); E. P. Wigner, *Am. J. Math.* **63**, 57 (1941).

see equation (3.21) below. In (3.16) it should be noted that the third projection quantum number is always the sum of the first two, since otherwise the coefficients vanish. For clarity we continue to write all three projection quantum numbers for the time being.

Some other useful symmetry relations, which can be derived from those given in (3.16a–c), are the following. Using (3.16c), (3.16b), and (3.16c) again, we find

$$C(j_1 j_2 j_3; m_1 m_2 m_3) = (-)^{j_2 + m_2} \left(\frac{2j_3 + 1}{2j_1 + 1}\right)^{\frac{1}{2}} C(j_3 j_2 j_1; -m_3, m_2, -m_1)$$

$$(3.17a)$$

If we use (3.16c), (3.16b), and (3.16a), the result is

$$C(j_1 j_2 j_3; m_1 m_2 m_3) = (-)^{j_1 - m_1} \left(\frac{2j_3 + 1}{2j_2 + 1}\right)^{\frac{1}{2}} C(j_3 j_1 j_2; m_3, -m_1, m_2)$$

$$(3.17b)$$

Using (3.17a), (3.16b), and (3.16a), we derive the result

$$C(j_1 j_2 j_3; m_1 m_2 m_3) = (-)^{j_2 + m_2} \left(\frac{2j_3 + 1}{2j_1 + 1}\right)^{\frac{1}{2}} C(j_2 j_3 j_1; -m_2 m_3 m_1)$$

$$(3.17c)$$

Again, the structure of these relations should be studied closely. A convenient description is the following: Whenever j_3 is interchanged with either j_1 or j_2, as in (3.16c) or (3.17a), the projection quantum number m_3 associated with it changes sign, and the projection quantum number associated with the angular momentum which is not interchanged with j_3 remains the same. The phase factor depends on the quantum numbers not involved in the interchange. The statistical weights in the square root involve j_3 (in the numerator) and the j-number which is interchanged with j_3 in the denominator. In (3.17b) and (3.17c) there are two transpositions of the j-numbers, and they appear as a cyclic permutation of j_1, j_2, j_3.

Wigner's closed expression [7] for the C-coefficients is

$$C(j_1 j_2 j_3; m_1 m_2 m_3) = \delta_{m_3, m_1 + m_2}$$

$$\times \left[(2j_3 + 1) \frac{(j_3 + j_1 - j_2)!(j_3 - j_1 + j_2)!(j_1 + j_2 - j_3)!(j_3 + m_3)!(j_3 - m_3)!}{(j_1 + j_2 + j_3 + 1)!(j_1 - m_1)!(j_1 + m_1)!(j_2 - m_2)!(j_2 + m_2)!} \right]^{\frac{1}{2}}$$

$$\times \sum_{\nu} \frac{(-)^{\nu + j_2 + m_2}}{\nu!} \frac{(j_2 + j_3 + m_1 - \nu)!(j_1 - m_1 + \nu)!}{(j_3 - j_1 + j_2 - \nu)!(j_3 + m_3 - \nu)!(\nu + j_1 - j_2 - m_3)!}$$

$$(3.18)$$

[7] E. P. Wigner, *Gruppentheorie*, Friedrich Vieweg und Sohn, Braunschweig, 1931.

The index ν assumes all integral values such that none of the factorial arguments are negative. In fact $1/(-n)! = 0$ for n a positive integer. All the arguments of the factorials are integers.

The derivation of the symmetry relations from this expression is fairly tedious, but may be greatly facilitated by using, instead, an expression of Racah's [5]

$$C(j_1 j_2 j_3; m_1 m_2 m_3) = \delta_{m_3, m_1 + m_2}$$

$$\times \left[(2j_3 + 1) \frac{(j_1 + j_2 - j_3)!(j_3 + j_1 - j_2)!(j_3 + j_2 - j_1)!}{(j_1 + j_2 + j_3 + 1)!} \right.$$

$$\left. \times (j_1 + m_1)!(j_1 - m_1)!(j_2 + m_2)!(j_2 - m_2)!(j_3 + m_3)!(j_3 - m_3)! \right]^{\frac{1}{2}}$$

$$\times \sum_\nu \frac{(-)^\nu}{\nu!} [(j_1 + j_2 - j_3 - \nu)!(j_1 - m_1 - \nu)!(j_2 + m_2 - \nu)!$$

$$\times (j_3 - j_2 + m_1 + \nu)!(j_3 - j_1 - m_2 + \nu)!]^{-1} \quad (3.19)$$

Once more, the integral index ν assumes only those values for which the factorial arguments are not negative, and again this is automatically achieved.

The three symmetry relations of (3.16) can now be easily derived. First we consider all of (3.19) except the sum over ν, or, in other words, the square root expression, since the conservation rule $\delta_{m_3, m_1 + m_2}$ is always satisfied. Inspection of this square root shows that it is invariant to the interchanges made in the three equations except that made in (3.16c), where the change is just the square root of the statistical weights $[(2j_3 + 1)/(2j_2 + 1)]^{\frac{1}{2}}$. The phase changes must then come from the sum over ν. Now it should be noticed that changing the signs of the projection quantum numbers in this sum, giving

$$\sum_\nu \frac{(-)^\nu}{\nu!} [(j_1 + j_2 - j_3 - \nu)!(j_1 + m_1 - \nu)!(j_2 - m_2 - \nu)!$$

$$\times (j_3 - j_2 - m_1 + \nu)!(j_3 - j_1 + m_2 + \nu)!]^{-1},$$

is equivalent to interchanging j_1 with j_2 and m_1 with m_2. This verifies that the phases in (3.16a) and (3.16b) must be the same. If the new summation index ν' is introduced by

$$\nu = j_1 + j_2 - j_3 - \nu'$$

this sum over ν in (3.19) becomes

$$(-)^{j_1 + j_2 - j_3} \sum_{\nu'} \frac{(-)^{\nu'}}{\nu'!} [(j_1 + j_2 - j_3 - \nu')!(j_3 - j_2 + m_1 + \nu')!$$

$$\times (j_3 - j_1 - m_2 + \nu')!(j_1 - m_1 - \nu')!(j_2 + m_2 - \nu')!]^{-1}$$

Except for the phase $(-)^{j_1+j_2-j_3}$, this is identical with the sum in (3.19). Therefore, changing the overall signs of the projection quantum numbers introduces the phase $(-)^{j_1+j_2-j_3}$ [in agreement with (3.16a)] as does the interchange of j_1 with j_2 and m_1 with m_2 [in agreement with (3.16b)].

To check the phase in the third symmetry relation we interchange j_2 with j_3, change m_2 to $-m_3$, and change m_3 to $-m_2$ in the sum and get

$$\sum_\nu \frac{(-)^\nu}{\nu!} [(j_1 + j_3 - j_2 - \nu)!(j_1 - m_1 - \nu)!(j_3 - m_3 - \nu)!$$

$$\times (j_2 - j_3 + m_1 + \nu)!(j_2 - j_1 + m_3 + \nu)!]^{-1}$$

The substitution

$$\nu = j_1 - m_1 - \nu'$$

leads to

$$(-)^{j_1-m_1} \sum_{\nu'} \frac{(-)^{\nu'}}{\nu'!} [(j_1 - m_1 - \nu')!(j_3 - j_2 + m_1 + \nu')!$$

$$\times (j_3 - j_1 - m_2 + \nu')!(j_2 - j_3 + j_1 - \nu')!(j_2 + m_2 - \nu')!]^{-1}$$

Comparison with the sum in (3.19) shows that interchange of j_2 and j_3 introduces the phase $(-)^{j_1-m_1}$ as well as the numerical factor $[(2j_3 + 1)/(2j_2 + 1)]^{\frac{1}{2}}$, in agreement with (3.16c). As has been shown above, the other symmetry relations (3.17) can be derived from these three. This establishes all the symmetry relations as a consequence of the explicit expression (3.19).

Racah [5] also introduced the so-called V-coefficients

$$C(j_1j_2j_3; m_1m_2m_3) = (-)^{j_3+m_3}(2j_3 + 1)^{\frac{1}{2}} V(j_1j_2j_3; m_1, m_2, -m_3) \quad (3.20)$$

which allow the symmetry relations to be rewritten in more balanced form. They now become

$$V(j_1j_2j_3; m_1m_2m_3) = (-)^{j_1+j_2+j_3} V(j_1j_2j_3; -m_1, -m_2, -m_3) \quad (3.21a)$$

$$= (-)^{j_1+j_2-j_3} V(j_2j_1j_3; m_2m_1m_3) \quad (3.21b)$$

$$= (-)^{j_1+j_2+j_3} V(j_1j_3j_2; m_1m_3m_2) \quad (3.21c)$$

$$= (-)^{j_1-j_2+j_3} V(j_3j_2j_1; m_3m_2m_1) \quad (3.21d)$$

$$= (-)^{2j_2} V(j_3j_1j_2; m_3m_1m_2) \quad (3.21e)$$

$$= (-)^{2j_3} V(j_2j_3j_1; m_2m_3m_1) \quad (3.21f)$$

The results (3.7) and (3.9) would be translated into

$$\sum_{m_1} V(j_1j_2j; m_1, m_2, -m_3) V(j_1j_2j'; m_1, m_2, -m_3) = \frac{\delta_{jj'}}{2j + 1}$$

$$(3.7')$$

$$\sum_j (2j + 1)\, V(j_1 j_2 j; m_1, m_2, -m_3)\, V(j_1 j_2 j; m_1', m_2', -m_3') = \delta_{m_1 m_1'}\, \delta_{m_2 m_2'}$$

(3.9′)

with $m_1 + m_2 = m_3$, $m_1' + m_2' = m_3'$.

Some properties of frequently encountered C-coefficients can be deduced from the symmetry relations. If the three angular momenta are integers, for example, the orbital angular momenta l_1, l_2, and l_3, and if all the projection quantum numbers vanish, (3.16a) gives

$$C(l_1 l_2 l_3; 000) = (-)^{l_1 + l_2 - l_3}\, C(l_1 l_2 l_3; 000)$$

Since l_3 is an integer, $(-)^{l_3} = (-)^{-l_3}$, and for this C-coefficient to be non-vanishing we require that $(-)^{l_1 + l_2 + l_3} = 1$, or $l_1 + l_2 + l_3$ must be even:

$$C(l_1 l_2 l_3; 000) = 0 \quad \text{unless} \quad l_1 + l_2 + l_3 \text{ is even} \tag{3.22}$$

This C-coefficient is called the parity C-coefficient since, as will be seen later, it contains the parity selection rule.

Next, consider the C-coefficient with $j_2 = 0$, and therefore $m_2 = 0$. From the conservation rules $\Delta(j_1 j_2 j_3)$ and $m_3 = m_1 + m_2$, we know that $C(j_1 0 j_3; m_1 0 m_3)$ vanishes unless $j_1 = j_3$ and $m_1 = m_3$. Further, the symmetry relation (3.17a) is

$$C(j_1 0 j_3; m_1 0 m_3) = \delta_{j_1 j_3}\, \delta_{m_1 m_3}\, C(j_3 0 j_1; -m_3, 0, -m_1)$$

The minus signs on the projection quantum numbers on the right can be changed by application of the first symmetry relation (3.16a),

$$C(j_1 0 j_3; m_1 0 m_3) = \delta_{j_1 j_3}\, \delta_{m_1 m_3}\, C(j_3 0 j_1; m_3 0 m_1)$$

$$= \delta_{j_1 j_3}\, \delta_{m_1 m_3}\, C(j_1 0 j_1; m_1 0 m_1)$$

This C-coefficient occurs for the case when no angular momentum is added to the angular momentum j_1, i.e.,

$$\psi_{j_1 m_1} = C(j_1 0 j_1; m_1 0 m_1)\, \psi_{j_1 m_1}$$

and so

$$C(j_1 0 j_3; m_1 0 m_3) = \delta_{j_1 j_3}\, \delta_{m_1 m_3} \tag{3.23}$$

By means of the symmetry relations and (3.23) we can evaluate any C-coefficient when any one of j_1, j_2, or j_3 is zero.

12. EVALUATION OF CLEBSCH–GORDAN COEFFICIENTS

Numerical values of the C-coefficients are not often used directly since a great deal of the analysis can usually be accomplished with the aid of the orthogonality relations (3.7) and the symmetry relations (3.16). However, on occasion numerical values are needed for the minimum j equal to some small number and the m-values having specific values. Examples will occur in Chapter X. Condon and Shortley [8] give tables of the coefficients most frequently needed, i.e., $C(j_1 j_2 j; m_1 m_2 m)$ for $j_2 = \frac{1}{2}, 1, \frac{3}{2}, 2$. Further tabulations are given by Saito and Morita for $j_2 = \frac{5}{2}$ and by Falkoff, Calladay, and Sells [9] for $j_2 = 3$. These tables are applicable, of course, when any one of the angular momenta has one of the above values, since the symmetry relations allow its transposition to the second place. A decimal tabulation due to Simon [10] is also available for cases where the angular momenta are all less than $\frac{9}{2}$.

The derivation of the explicit formulae (3.18) and (3.19) is rather lengthy. We can devise a rather simple method which is capable of giving equivalent results except that a phase depending on j_3 is left arbitrary. From what has been said so far, this phase is not defined, and (3.18) or (3.19) result when a specific phase convention has been adopted. The details of the procedure by means of which this may be done are discussed in Appendix I.

The simplified derivation is based on a recurrence formula for the C-coefficients which we now derive. If we combine two angular momenta to form a resultant $\mathbf{J} = \mathbf{J}_1 + \mathbf{J}_2$, then, according to section 10, the eigenfunctions of $\mathbf{J}^2$ and J_z, are

$$\psi_{JM} = \sum_m C(j_1 j_2 J; m, M-m)\, \psi_{j_1 m}\, \psi_{j_2\, M-m} \qquad (3.24)$$

We now apply $\mathbf{J}^2$ to both sides of this equation, using on the right side

$$\mathbf{J}^2 = \mathbf{J}_1^2 + \mathbf{J}_2^2 + 2\mathbf{J}_1 \cdot \mathbf{J}_2$$

For any two vectors $\mathbf{A}$ and $\mathbf{B}$,

$$\mathbf{A} \cdot \mathbf{B} = A_z B_z + \tfrac{1}{2}(A_+ B_- + A_- B_+)$$

$$A_\pm = A_x \pm i A_y \quad \text{and} \quad B_\pm = B_x \pm i B_y \qquad (3.25)$$

In our case, therefore, $\mathbf{J}^2$ can be written as

$$\mathbf{J}^2 = \mathbf{J}_1^2 + \mathbf{J}_2^2 + J_{1+}J_{2-} + J_{1-}J_{2+} + 2J_{1z}J_{2z} \qquad (3.26)$$

[8] Condon and Shortley, reference 4, pp. 70–76.

[9] R. Saito and M. Morita, *Prog. Theoret. Phys. Japan* **13**, 540 (1955). D. L. Falkoff, C. S. Calladay, and R. E. Sells, *Can. J. Phys.* **30**, 253 (1952). See also B. J. Sears and M. G. Radtke, *Chalk River Rept. TPI-75* (August 1954).

[10] A. Simon, Numerical Tables of the Clebsch–Gordan Coefficients, *Oak Ridge Nat. Lab. Rept. 1718*.

Application of $\mathbf{J}^2$ to (3.24) gives

$$J(J+1) \sum_m C(j_1 j_2 J; m, M-m) \, \psi_{j_1 m} \psi_{j_2 M-m}$$
$$= \sum_m C(j_1 j_2 J; m, M-m)$$
$$\times [j_1(j_1+1) + j_2(j_2+1) + 2m(M-m)] \psi_{j_1 m} \psi_{j_2 M-m}$$
$$+ \sum_m C(j_1 j_2 J; m, M-m) [J_{1+}J_{2-} + J_{1-}J_{2+}] \psi_{j_1 m} \psi_{j_2 M-m}$$

In the last line we use the matrix elements of the raising and lowering operators as given in (2.28) to yield the result

$$\sum_m [J(J+1) - j_1(j_1+1) - j_2(j_2+1) - 2m(M-m)]$$
$$\times C(j_1 j_2 J; m, M-m) \, \psi_{j_1 m} \psi_{j_2 M-m} = \sum_m C(j_1 j_2 J; m, M-m)$$
$$\times [(j_1-m)(j_1+m+1)(j_2+M-m)(j_2-M+m+1)]^{\frac{1}{2}} \psi_{j_1 m+1} \psi_{j_2 M-m-1}$$
$$+ \sum_m C(j_1 j_2 J; m, M-m) [(j_1+m)(j_1-m+1)(j_2-M+m)$$
$$\times (j_2+M-m+1)]^{\frac{1}{2}} \psi_{j_1 m-1} \psi_{j_2 M-m+1}$$

On the right side we introduce the new summation indices, $m' = m + 1$ in the first sum and $m'' = m - 1$ in the second sum so that the last equation may now be written

$$\sum_m [J(J+1) - j_1(j_1+1) - j_2(j_2+1) - 2m(M-m)]$$
$$\times C(j_1 j_2 J; m, M-m) \, \psi_{j_1 m} \psi_{j_2 M-m} = \sum_{m'} [(j_1 - m' + 1)(j_1 + m')$$
$$\times (j_2 + M - m' + 1)(j_2 - M + m')]^{\frac{1}{2}} C(j_1 j_2 J; m'-1, M-m'+1)$$
$$\times \psi_{j_1 m'} \psi_{j_2 M-m'} + \sum_{m''} [(j_1 + m'' + 1)(j_1 - m'')(j_2 - M + m'' + 1)$$
$$\times (j_2 + M - m'')]^{\frac{1}{2}} C(j_1 j_2 J; m''+1, M-m''-1) \psi_{j_1 m''} \psi_{j_2 M-m''}$$

The changes in the limits are taken care of by the rules for the vanishing of the C-coefficients. The primes and double primes can be dropped, and, since the eigenfunctions are linearly independent, we equate the coefficients of $\psi_{j_1 m} \psi_{j_2 M-m}$ on each side to obtain

$$[J(J+1) - j_1(j_1+1) - j_2(j_2+1) - 2m(M-m)] C(j_1 j_2 J; m, M-m)$$
$$= [(j_1 - m + 1)(j_1 + m)(j_2 + M - m + 1)(j_2 - M + m)]^{\frac{1}{2}}$$
$$\times C(j_1 j_2 J; m-1, M-m+1)$$
$$+ [(j_1 + m + 1)(j_1 - m)(j_2 - M + m + 1)(j_2 + M - m)]^{\frac{1}{2}}$$
$$\times C(j_1 j_2 J; m+1, M-m-1) \quad (3.27)$$

This is a recurrence relation for the Clebsch–Gordan coefficients in the projection quantum numbers for fixed values of j_1, j_2, J, and M. The usual practice is to write a matrix for the C-coefficients with rows labeled by J values ($|j_1 - j_2| \leq J \leq j_1 + j_2$), the first row being $J = j_1 + j_2$, and columns labeled by m_2 values ($-j_2 \leq m_2 \leq j_2$), the first column being $m_2 = j_2$. Reference may be made to Appendix I, Tables I.1 and I.2. The recurrence relation (3.27) works across the rows of this matrix. Given one of the elements, for instance, that one at the extreme left end of a row, the remainder of the elements in the row can be generated.

As an example of the use of the recurrence formula (3.27), consider the case $j_2 = 1$ and $J = j_1 + 1$, i.e., the top row of the matrix for $j_2 = 1$. For the sake of brevity, we introduce

$$x_\mu = C(j_1 1\, j_1+1; M-\mu, \mu)$$

where $\mu = M - m_1$ is the second projection quantum number. There are three recurrence relations, one for each value of $\mu = 1, 0, -1$:

$\mu = 1$: $\quad 2(j_1 - M + 1)\, x_1 = [2(j_1 + M)(j_1 - M + 1)]^{\frac{1}{2}}\, x_0$

$\mu = 0$: $\quad 2j_1\, x_1 = [2(j_1 - M + 1)(j_1 + M)]^{\frac{1}{2}}\, x_1$

$$+ [2(j_1 + M + 1)(j_1 - M)]^{\frac{1}{2}}\, x_{-1}$$

$\mu = -1$: $\quad 2(j_1 + M + 1)\, x_{-1} = [2(j_1 - M)(j_1 + M + 1)]^{\frac{1}{2}}\, x_0$

Since these are homogeneous equations, only ratios can be determined; for example,

$$\frac{x_1}{x_0} = \left[\frac{(j_1 + M)}{2(j_1 - M + 1)} \right]^{\frac{1}{2}} \quad \text{and} \quad \frac{x_{-1}}{x_0} = \left[\frac{j_1 - M}{2(j_1 + M + 1)} \right]^{\frac{1}{2}}$$

When the orthogonality condition (3.7) is applied, we get

$$x_1^2 + x_0^2 + x_{-1}^2 = 1$$

which fixes all the x_μ as far as magnitude and relative phase is concerned. Still, the overall phase remains undetermined since this equation together with the above ratios determines only the square of one of the coefficients, for example

$$x_0^2 = \frac{(j_1 - M + 1)(j_1 + M + 1)}{(j_1 + 1)(2j_1 + 1)}$$

In section 10, however, we adopted the convention that $C(j_1 j_1\, j_1 + j_2; j_1 j_2)$ = 1. In the case at hand, this choice means that the positive square root is to be taken since x_1 must be 1 for $M = j_1 + 1$. The first row of the 3-by-3 matrix $C(j_1 1 J; M - \mu, \mu)$ is now completely determined:

$$x_1 = \left[\frac{(j_1 + M)(J_1 + M + 1)}{(2j_1 + 1)(2j_1 + 2)} \right]^{\frac{1}{2}} = C(j_1 1\, j_1+1; M-1, 1)$$

$$x_0 = \left[\frac{(j_1 - M + 1)(j_1 + M + 1)}{(2j_1 + 1)(j_1 + 1)} \right]^{\frac{1}{2}} = C(j_1 1\, j_1+1; M0)$$

$$x_{-1} = \left[\frac{(j_1 - M)(j_1 - M + 1)}{(2j_1 + 1)(2j_1 + 2)} \right]^{\frac{1}{2}} = C(j_1 1\, j_1+1; M+1, -1) \quad (3.28)$$

In obtaining the second and third rows, an ambiguity in the overall sign of each row is once more found. The convention used above for the first row does not apply here. These questions of phase convention are, of course, a general characteristic of unitary transformations. A change in overall phase of any row or column of a unitary matrix (i.e., multiplication by a common factor $e^{i\delta}$, where δ is real) does not alter the unitary character. This can be readily seen by inspection of the orthogonality relations (3.7) and (3.9). Alternatively, we can express the same fact by observing that the wave functions ψ_{JM} can always be multiplied by a (real) phase which depends on J. The results expressed in (3.18) or (3.19) corresponding to a definite choice of this phase. In an ad hoc manner we can simplify this question by simply saying that the C-coefficients for $m_2 = -j_2$ shall be non-negative. From (3.19) these coefficients are

$$C(j_1 j_2 j_3; m_1, -j_2)$$

$$= \left[\frac{(2j_3+1)(j_3+j_1-j_2)!(2j_2)!}{(j_1+j_2-j_3)!(j_3-j_1+j_2)!(j_1+j_2+j_3+1)!} \frac{(j_1+m_1)!(j_3+j_2-m_1)!}{(j_1-m_1)!(j_3-j_2+m_1)!} \right]^{\frac{1}{2}}$$

$$(3.29)$$

From (3.18) or (3.19) it is possible to obtain any C-coefficient in explicit form. The cases $j_2 = \frac{1}{2}$ and $j_2 = 1$ are especially important, and the results are presented in Appendix I—Tables I.1 and I.2.

A result of particular interest for later applications can be derived at once from (3.27). In that result we set $j_1 = j_2 = L$ and $J = \nu$ (to accord with a notation used subsequently). Also we set $m = M = 0$. Then, using (3.16a), we find at once

$$C(LL\nu; 00) = \frac{2L(L + 1)}{\nu(\nu + 1) - 2L(L + 1)} C(LL\nu; 1, -1) \quad (3.30)$$

where ν is even. Otherwise the coefficient $C(LL\nu; 00)$ vanishes, see (3.22). More generally we obtain in the same way

$$C(LL'\nu; 00) = \frac{[L(L + 1) L'(L' + 1)]^{\frac{1}{2}}}{\nu(\nu + 1) - L(L + 1) - L'(L' + 1)}$$
$$\times [1 + (-)^{L+L'-\nu}] C(LL'\nu; 1, -1) \quad (3.31)$$

which shows explicitly that $C(LL'\nu; 00)$ vanishes if $L + L' + \nu$ is odd.
Another useful result, first given by Racah,[5] is

$$C(L_1 L_2 L_3; 00) = (-)^{\frac{1}{2}(L_1+L_2-L_3)} \left(\frac{2L_3 + 1}{L_1 + L_2 + L_3 + 1}\right)^{\frac{1}{2}}$$
$$\times \frac{\tau(L_1 + L_2 + L_3)}{\tau(L_1 + L_2 - L_3)\tau(L_1 - L_2 + L_3)\tau(-L_1 + L_2 + L_3)} \quad (3.32)$$

where

$$\tau(x) = \frac{(\frac{1}{2}x)!}{\sqrt{x!}}$$

provided $L_1 + L_2 + L_3$ is an even integer. The results given in (3.30) and (3.32) are important for angular distributions (section 34) and angular correlations (section 33) of alpha particles or other spin zero particles and photons (spin 1 particles).

IV. TRANSFORMATION PROPERTIES UNDER ROTATIONS

13. MATRIX REPRESENTATIONS OF THE ROTATION OPERATORS

The explicit functions ψ_{jm} which diagonalize the square and one component of the angular momentum, with eigenvalues $j(j + 1)$ and m, respectively, presuppose a definite choice of the quantization axis. This is so because m is the eigenvalue of the angular momentum component on that axis. It is often necessary to consider rotations of the axis of quantization, and this chapter will be devoted to a study of the transformation properties of the angular momentum eigenfunctions under these rotations. This study also provides a very important classification of the operators which represent interactions in quantum mechanics.

The eigenvalue of the square of the angular momentum is unchanged by a rotation, since $\mathbf{J}^2$ commutes with the rotation operator $\exp(-i\theta\mathbf{n}\cdot\mathbf{J})$, the coordinate system being rotated through an angle θ about the direction $\mathbf{n}$. Expressed formally,

$$[\mathbf{J}^2, e^{-i\theta(\mathbf{n}\cdot\mathbf{J})}] = \left[\mathbf{J}^2, \sum_\nu \frac{1}{\nu!}(-i\theta\mathbf{n}\cdot\mathbf{J})^\nu\right] = \sum_\nu \frac{1}{\nu!}(-i\theta)^\nu[\mathbf{J}^2, (\mathbf{n}\cdot\mathbf{J})^\nu] = 0.$$

In other words, $\mathbf{J}^2$ commutes with each term of the series expansion of the rotation operator. Hence the rotated function, given in (2.8),

$$R\psi_{jm} = e^{-i\theta(\mathbf{n}\cdot\mathbf{J})}\psi_{jm} \tag{4.1}$$

is an eigenfunction of $\mathbf{J}^2$ with unchanged eigenvalue $j(j + 1)$. Explicitly,

$$\mathbf{J}^2(e^{-i\theta(\mathbf{n}\cdot\mathbf{J})}\psi_{jm}) = e^{-i\theta(\mathbf{n}\cdot\mathbf{J})}(\mathbf{J}^2\psi_{jm}) = j(j + 1)(e^{-i\theta(\mathbf{n}\cdot\mathbf{J})}\psi_{jm})$$

This result, quite naturally, expresses the well-known fact that the total angular momentum has nothing to do with a particular direction or a particular coordinate system.

The rotated function $R\psi_{jm}$ does not diagonalize J_z, in general, and is a superposition of eigenfunctions $\psi_{jm'}$ with different projection quantum numbers m' but the same total angular momentum j.

$$R\psi_{jm} = \sum_{m'} (jm'|e^{-i\theta(\mathbf{n}\cdot\mathbf{J})}|jm)\,\psi_{jm'} \tag{4.2}$$

48

An especially simple case is that of a rotation about the quantization axis itself, $\mathbf{n} \cdot \mathbf{J} = J_z$. Then

$$(jm' | e^{-i\theta(\mathbf{n} \cdot \mathbf{J})} | jm) = \delta_{m'm}\, e^{-im\theta}$$

and the projection quantum number remains the same, the only change being one of phase according to

$$R\psi_{jm} = e^{-im\theta}\, \psi_{jm} \qquad (4.3)$$

This is the situation with the rigid rotator in a plane, discussed in section 4. The normalized wave function is

$$\psi_{jm}(\phi) = \frac{1}{(2\pi)^{\frac{1}{2}}}\, e^{im\phi}$$

where ϕ is the orientation of the rotator with respect to a fixed line OA and m is the (integral) eigenvalue of the angular momentum component.

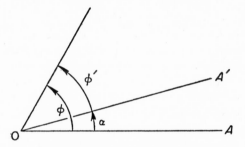

Fig. 2. Rotation angles in the plane.

Here $\mathbf{J}^2 = J_z^2$. If the reference line is rotated through an angle α to OA', as in Fig. 2, the rotated wave function is

$$R\psi_{jm}(\phi) = \frac{1}{(2\pi)^{\frac{1}{2}}}\, e^{im\phi'} = \frac{1}{(2\pi)^{\frac{1}{2}}}\, e^{im(\phi-\alpha)}$$

since $\phi' = \phi - \alpha$. Therefore

$$R\psi_{jm}(\phi) = e^{-im\alpha}\, \psi_{jm}(\phi)$$

which agrees with (4.3).

Returning to the general problem, which is to determine the matrix representation $(jm' | \exp(-i\theta\mathbf{n} \cdot \mathbf{J}) | jm)$ of the rotation operator $R = \exp(-i\theta\mathbf{n} \cdot \mathbf{J})$, we first review some elementary facts about rotations of coordinate systems. It is well known that three parameters are needed to specify a rotation. These may be the three components of the vector $\theta\mathbf{n}$, where θ gives the magnitude and $\mathbf{n}$ the direction of the rotation.

The most useful description is, however, in terms of the Euler angles,[1] which we shall call α, β, γ. These are defined by performing the rotation of the coordinate system in three steps, the original coordinate axis being x, y, z:

1. A rotation is made about the z-axis through an angle α; the new coordinate axes are x', y', z'; see Fig. 3a.

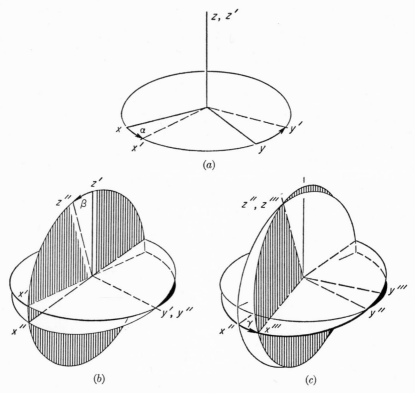

Fig. 3. The Euler angles α, β and γ and the three Euler rotations which carry the initial (x, y, z) coordinate system into the final (x''', y''', z''') coordinate system.

2. A rotation is made about the y'-axis through an angle β; the new coordinate axes are x'', y'', z'', as in Fig. 3b.

[1] See, for example, H. Goldstein, *Classical Mechanics*, p. 107, Addison-Wesley, Cambridge, 1953. The author calls attention to the fact that many different choices of Euler angles are used in the literature. In fact, the Euler angles defined here differ from those used by Goldstein in that his second rotation is about the x'-axis, rather than about the y'-axis. Note also that Wigner (*Gruppentheorie*, Friedrich Vieweg und Sohn, Braunschweig, 1931) adopts a different choice in that he uses a left-handed coordinate system.

3. A rotation is made about the z''-axis through an angle γ; the new coordinate axes are x''', y''', z'''; see Fig. 3c.

In the diagrams of Fig. 3 the Euler rotations have been illustrated as positive rotations. A positive rotation is defined here as one that advances a right-hand screw along the axis of rotation.

The rotation operator in (4.1) is now a product of three operators

$$e^{-i\theta(\mathbf{n}\cdot\mathbf{J})} = R(\theta\mathbf{n}) = R_\gamma R_\beta R_\alpha \tag{4.4}$$

with R_α operating on the wave function first. Obviously, the subscripts are the Euler angles of rotation. But R_α, R_β, and R_γ are successive rotations about the z, y', and z'' axes, respectively; therefore,

$$R = R_\gamma R_\beta R_\alpha = e^{-i\gamma J_{z''}} e^{-i\beta J_{y'}} e^{-i\alpha J_z} \tag{4.5}$$

Here $J_{y'}$ and $J_{z''}$ are the components of $\mathbf{J}$ along the y' and z'' axes, respectively. According to section 2, however, a unitary transformation U transforms an operator Ω into $U\Omega U^{-1}$. Thus $R_\beta = \exp(-i\beta J_{y'})$ is the transform of $\exp(-i\beta J_y)$ under the previous rotation $R_\alpha = \exp(-i\alpha J_z)$ which carried the y-axis into the y'-axis; and $R_\gamma = \exp(-i\gamma J_{z''})$ is the transform of $\exp(-i\gamma J_{z'})$ under the previous rotation $R_\beta = \exp(-i\beta J_{y'})$ which carried z' into z''. Accordingly, to express R_β in terms of the original coordinate system we use $U = R_\alpha$ and

$$e^{-i\beta J_{y'}} = e^{-i\alpha J_z} e^{-i\beta J_y} e^{i\alpha J_z} \tag{4.6a}$$

In the same way

$$e^{-i\gamma J_{z''}} = e^{-i\beta J_{y'}} e^{-i\gamma J_{z'}} e^{i\beta J_{y'}} \tag{4.6b}$$

which expresses R_γ in the coordinate system that results after the R_α rotation is performed. If we use (4.6b), the rotation operator R in (4.5) is now

$$R = e^{-i\beta J_{y'}} e^{-i\gamma J_{z'}} e^{-i\alpha J_z}$$

Applying (4.6a) and a similar equation for $\exp(-i\gamma J_{z'})$ gives, finally,

$$R = e^{-i\alpha J_z} e^{-i\beta J_y} e^{-i\gamma J_z} \tag{4.7}$$

Equation (4.5) expresses the fact that the rotation R is carried out by the three successive Euler rotations: The first is a rotation α along the z-axis, the second a rotation β about the y'-axis, and the third a rotation γ about the z''-axis. Thus, the axes of rotation are coordinate axes of different coordinate systems, namely the coordinate system obtained by the previous rotation. Equation (4.7) says that these rotations may all be carried out in the *same* coordinate system if the order of the rotations is inverted. That is, the rotation R of the coordinate system may also be performed in this way: First a rotation γ is made about the z-axis, then a rotation β is made about the y-axis, and, finally, a rotation α is made about the z-axis.

The dependence on α and γ of the matrix representation of the rotation operator R can now be determined very simply. Following Wigner's convention,[2] we call this matrix $D^j_{m'm}(\alpha\beta\gamma)$ so that equation (4.2) becomes

$$R\psi_{jm} = \sum_{m'} D^j_{m'm}(\alpha\beta\gamma)\,\psi_{jm'} \tag{4.8}$$

If we use equation (4.7) for R, the elements of the rotation matrix $D^j_{m'm}(\alpha\beta\gamma)$ are

$$D^j_{m'm}(\alpha\beta\gamma) = (jm'|e^{-i\alpha J_z}\,e^{-i\beta J_y}\,e^{-i\gamma J_z}|jm)$$

$$= (\psi_{jm'},\, e^{-i\alpha J_z}\,e^{-i\beta J_y}\,e^{-i\gamma J_z}\,\psi_{jm}) \tag{4.9}$$

Letting $\exp(-i\gamma J_z)$ operate on the right and $\exp(-i\alpha J_z)$ operate on the left, using equation (1.2), for $[\exp(-i\alpha J_z)]^+ = \exp(i\alpha J_z)$ and also equations (1.1a) and (1.1b), we find that

$$D^j_{m'm}(\alpha\beta\gamma) = e^{-im'\alpha}\,(jm'|e^{-i\beta J_y}|jm)\,e^{-im\gamma} \tag{4.10}$$

Of course, $\exp(-i\beta J_y)$ is not diagonal in this representation. Calling its matrix elements $d^j_{m'm}(\beta)$, that is

$$d^j_{m'm}(\beta) = (jm'|e^{-i\beta J_y}|jm) \tag{4.11}$$

the (factored) dependence of $D^j_{m'm}(\alpha\beta\gamma)$ on the Euler angles is

$$D^j_{m'm}(\alpha\beta\gamma) = e^{-im'\alpha}\,d^j_{m'm}(\beta)\,e^{-im\gamma} \tag{4.12}$$

Wigner [2] has given the following expression for $d^j_{m'm}(\beta)$:

$$d^j_{m'm}(\beta) = [(j+m)!(j-m)!(j+m')!(j-m')!]^{\frac{1}{2}}$$

$$\times \sum_{\varkappa} \frac{(-)^{\varkappa}}{(j-m'-\varkappa)!(j+m-\varkappa)!(\varkappa+m'-m)!\varkappa!}$$

$$\times \left(\cos\frac{\beta}{2}\right)^{2j+m-m'-2\varkappa} \left(-\sin\frac{\beta}{2}\right)^{m'-m+2\varkappa} \tag{4.13}$$

[2] Pages 180 and 232 of Wigner, reference 1. The $D^j_{m'm}$ defined here is related to Wigner's $\mathfrak{D}^{(j)}_{m'm}$ by a change in sign in m' and m:

$$D^j_{m'm} = \mathfrak{D}^{(j)}_{-m',-m}$$

This is in accord with the convention of G. Goertzel, *Phys. Rev.* **70**, 897 (1946); G. Racah, **84**, 910 (1951); and with J. Schwinger, On Angular Momentum, *Nuclear Development Corp. of America Rept. NYO-3071* (January 26, 1952).

where the sum is over the values of the integer $\varkappa$ for which the factorial arguments are greater than or equal to zero. The result (4.13) is derived in Appendix II.

Another way of writing (4.13) is

$$d^j_{m'm}(\beta) = \left[\frac{(j-m)!(j+m')!}{(j+m)!(j-m')!}\right]^{\frac{1}{2}} \frac{\left(\cos\dfrac{\beta}{2}\right)^{2j+m-m'}\left(-\sin\dfrac{\beta}{2}\right)^{m'-m}}{(m'-m)!}$$

$$\times\, _2F_1\left(m'-j,\ -m-j;\ m'-m+1;\ -\tan^2\frac{\beta}{2}\right); \qquad m' \geq m \quad (4.14)$$

The function $_2F_1(a, b; c; z)$ occurring in (4.15) is the hypergeometric function,[3] which is symmetric in a and b. For $|z| < 1$ and c not zero or a negative integer, the following series expansion exists

$$_2F_1(a, b; c; z) = 1 + \frac{ab}{c}z + \frac{1}{2!}\frac{a(a+1)\,b(b+1)}{c(c+1)}z^2 + \cdots$$

Since a or b is, in our case, a negative integer, the series terminates and is valid for all z. In this case $F(-n, b; c; x)$ are Jacobi polynomials[4] of degree n and are orthogonal on the interval $0 \leq x \leq 1$ with respect to the weighting factor $x^{c-1}(1-x)^{b-c-n}$. In (4.14) the hypergeometric function is a Jacobi polynomial of degree $|m'-j|$ or $|m+j|$, whichever is smaller.

Equation (4.14) is valid only for $m' \geq m$; values for $m > m'$ may be obtained from (4.13) or from the unitary property

$$d^j_{m'm}(\beta) = d^j_{mm'}(-\beta) \qquad (4.15)$$

That is, the inverse rotation to $\exp(-i\beta J_y)$ is $\exp(i\beta J_y)$, a rotation about y through $-\beta$. But a rotation is a unitary transformation, and the inverse operator is the Hermitian adjoint. Thus, in finding $d_{mm'}(-\beta)$ we have transposed the indices and taken the complex conjugate of $d^*_{mm'}(\beta)$. The last operation is unnecessary, however, since inspection of (4.13) and (4.14) reveals that $d^j_{m'm}(\beta)$ is real. Now a replacement of β with $-\beta$ in (4.14) leads only to a phase change $(-)^{m'-m+2\varkappa}$ in the sum over $\varkappa$. But $\varkappa$ is an integer and $(-)^{2\varkappa} = 1$, so that changing the sign of β is equivalent to an overall phase change of $(-)^{m'-m}$,

$$d^j_{m'm}(-\beta) = (-)^{m'-m}\, d^j_{m'm}(\beta) \qquad (4.16)$$

[3] Bateman Manuscript Project, *Higher Transcendental Functions*, Vol. 1, Chapter II, McGraw-Hill Book Co., New York, 1953.

[4] Bateman Manuscript Project, *Higher Transcendental Functions*, Vol. 2, Chapter X, McGraw-Hill Book Co., New York, 1953.

Application of this result to (4.15) leads to this very useful relation

$$d^j_{m'm}(\beta) = (-)^{m'-m}\, d^j_{mm'}(\beta) \tag{4.17}$$

This result may be used to supplement equation (4.14) for $d^j_{m'm}(\beta)$, which is valid for $m' \geq m$. For, if the first index on the left side of (4.17) is smaller than the second ($m' < m$), the converse situation exists on the right side.

Another important relation can be derived from inspection of (4.13) or (4.14). It states that $d^j_{m'm}(\beta)$ is invariant to an interchange of m' and m accompanied by a change in sign of both m' and m,

$$d^j_{m'm}(\beta) = d^j_{-m,-m'}(\beta) \tag{4.18}$$

Coupled with (4.17), this means that

$$d^j_{m'm}(\beta) = (-)^{m'-m}\, d^j_{-m',-m}(\beta) \tag{4.19}$$

We have made use of the fact [equation (4.5)] that any rotation of the coordinate system can be carried out by three successive Euler rotations. But the inverse of a product of operators is the product of the inverse operators in inverse order

$$R^{-1} = e^{i\gamma J_z}\, e^{i\beta J_y}\, e^{i\alpha J_z} \tag{4.20}$$

Thus the inverse rotation is accomplished by performing the rotations through negative angles (about the same axes) but in opposite order to (4.7). The unitary property of the rotation operator R means that the matrix elements of R^{-1} are identical with those of the Hermitian adjoint

$$(jm'|R^{-1}|jm) = (jm|R|jm')^*$$

Using (4.20), this means that

$$D^j_{m'm}(-\gamma, -\beta, -\alpha) = D^{j*}_{mm'}(\alpha\beta\gamma) \tag{4.21}$$

This is a generalization of equations (4.15) and (4.17) for $d^j_{m'm}(\beta)$ and can easily be derived from (4.12) with the aid of (4.15):

$$D^j_{m'm}(-\gamma, -\beta, -\alpha) = e^{im'\gamma}\, d^j_{m'm}(-\beta)\, e^{im\alpha}$$

$$= e^{im\alpha}\, d^j_{mm'}(\beta)\, e^{im'\gamma} = D^{j*}_{mm'}(\alpha\beta\gamma)$$

The generalization of equation (4.19) is obtained in the following way

$$D^{j*}_{m'm}(\alpha\beta\gamma) = e^{im'\alpha}\, d_{m'm}(\beta)\, e^{im\gamma}$$

$$= (-)^{m'-m}\, e^{im'\alpha}\, d_{-m',-m}(\beta)\, e^{im\gamma}$$

where we have used equation (4.19); therefore,

$$D^{j*}_{m'm}(\alpha\beta\gamma) = (-)^{m'-m}\, D^j_{-m',-m}(\alpha\beta\gamma) \tag{4.22}$$

After this discussion of the formal properties it is of interest to mention the physical significance of the rotation matrices $D^j(\alpha\beta\gamma)$. For example, Wigner has shown that they are the wave functions of the symmetric top.[5] A symmetric top is simply a rigid body with two principal moments of inertia equal. It is almost evident that the D^j-matrices should have this property since they describe the effect of rotating an axially symmetric system as a rigid body. The Schrödinger equation for the symmetric top is

$$T \, \psi_{LKM}(\alpha\beta\gamma) = E \, \psi_{LKM}(\alpha\beta\gamma)$$

where the kinetic energy operator is

$$T = \frac{\hbar^2}{2I}(L_a^2 + L_b^2) + \frac{\hbar^2}{2I'}(L_c^2)$$

In this case the Euler angles $\alpha\beta\gamma$ define the rotation which carries a fixed coordinate system into the principal axes of inertia a, b, c fixed in the body. The components of the angular momentum operators along the principal axes are L_a, L_b, L_c. The normalized wave functions are, see (4.60) below,

$$\psi_{LKM}(\alpha\beta\gamma) = \left(\frac{2L+1}{8\pi^2}\right)^{\frac{1}{2}} D^L_{-K,-M}(\alpha\beta\gamma)$$

As usual, the square and z-component (space-fixed axis) of the angular momentum are diagonal with eigenvalues $L(L+1)$ and M, respectively. In addition, the component along the body-fixed symmetry axis is diagonal, and its eigenvalue [6] is K. Both K and M can assume integral values from $-L$ to L. The energy levels of the symmetric top are

$$E = \frac{1}{2I} L(L+1)\hbar^2 + \frac{1}{2}\left(\frac{1}{I'} - \frac{1}{I}\right) K^2\hbar^2 \qquad (4.23)$$

which follows from the fact that

$$T = \frac{\hbar^2}{2I} \mathbf{L}^2 + \frac{\hbar^2}{2}\left(\frac{1}{I'} - \frac{1}{I}\right) \mathbf{L}_c^2$$

where $\mathbf{L}^2 = L_a^2 + L_b^2 + L_c^2$. Each operator in T is separately diagonal.

[5] Page 231 of Wigner, reference 1. The wave functions for the symmetric top were first obtained by F. Reiche and H. Rademacher [Z. Physik 39, 444 (1926)] and independently by R. L. de Kronig and I. I. Rabi [Phys. Rev. 29, 262 (1927)].

[6] This component is also a constant of motion in the classical solution, where the motion consists of a rotation of the body about its symmetry axis (with constant angular velocity) plus a constant precession of the symmetry axis about the direction of the angular momentum.

These solutions describe the rotational motion of many molecules, and are also used in various nuclear models, notably the collective [7] model. In all these applications the description is based on a separation of the motion into two parts, one being the motion of the whole system as a rigid body and the other a more complicated internal or particle motion.

Another application concerns the effect of an arbitrary magnetic field $H(\mathbf{r}, t)$ on the orientation of the angular momentum of a system.[8,9] For example, Majorana has shown that the sudden application of a magnetic field is equivalent to a rotation of the angular momentum of the system. Consider a beam of particles of intrinsic angular momentum j, and suppose that we have a device which permits only the passage of particles whose z-component of spin is $m = j$. Then we can do a polarizer-analyzer experiment in the following way. An unpolarized beam of particles approaches the polarizer along the y-axis; see Fig. 4. The particles that

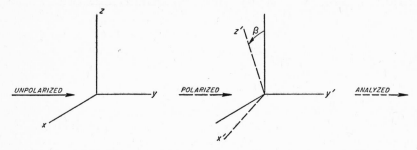

Fig. 4. The left-hand figure illustrates schematically an incident unpolarized beam which becomes polarized along the z-axis. The analyzer is rotated through an angle β about the y-axis.

pass the plane of the polarizer (the xz-plane) are all polarized with $m = j$. The analyzer is the same sort of device as the polarizer and has been rotated through an angle β about the y-axis. We want to know how many particles are passed by the analyzer, i.e., how many have maximum projection along the new quantization axis? The answer to this is to be found in the connection between the original eigenfunctions ψ_{jm} and the eigenfunctions with respect to the rotated quantization axis $R(0, \beta, 0)\,\psi_{jm}$ which is given in equation (4.8).

$$R\psi_{jj} = \sum_{m'} D^{j*}_{jm'}(0, \beta, 0)\,\psi_{jm'}$$

The number of particles that have a projection on the new quantization

[7] A. Bohr and B. R. Mottelson, *Danske Videnskab. Selskab, Math.-fys. Medd.* **27**, no. 16 (1953).

[8] E. Majorana, *Nuovo cimento* **9**, 43 (1932).

[9] F. Bloch and I. I. Rabi, *Revs. Mod. Phys.* **17**, 237 (1945).

axis of m' is $|D_{jm'}^j(0, \beta, 0)|^2$. Therefore the number that the analyzer allows to pass is $|D_{jj}^j(0, \beta, 0)|^2 = [d_{jj}^j(\beta)]^2$.

The properties of the rotation matrices can be discussed in very general terms using the principles of group theory.[10-12] A set of elements constitutes a group if there is an associative multiplication (within the group) such that

1. The product of any two elements of the group is also in the group.
2. There is an identity element.
3. The inverse of every element exists.

Actually (2) is implied by (1) and (3). For example, the set of three-dimensional rotations forms a group. Any two rotations performed successively is equivalent to some other single rotation. The identity element is the rotation through null angles, and the inverse of any rotation is the one about the same axis through the negative angle. Let $A, B, C, \cdots$ indicate the group elements. These letters serve to distinguish the various possible rotations. If there exists a set of matrices $D(A)$, $D(B), D(C), \cdots$ in one-to-one (or n-to-one) correspondence with and possessing the same properties as the group, then these matrices are said to constitute a one-to-one (or n-to-one) representation of the group. The $D_{m'm}^j(\alpha\beta\gamma)$ are the elements of matrices which, for each j, constitute a $2j+1$-dimensional representation of the 3-dimensional rotation group; see Appendix II.

14. THE CLEBSCH–GORDAN SERIES

The discussion of the previous section on the general properties of the rotation matrices $D^j(\alpha\beta\gamma)$ can now be applied to the derivation of some very useful results. In Chapter III we treated the coupling of two angular momenta j_1 and j_2 to give a resultant j. The connection between the uncoupled and coupled representations is given, for example, by equation (3.8),

$$\psi_{j_1m_1}\psi_{j_2m_2} = \sum_j C(j_1j_2j; m_1m_2)\psi_{j\ m_1+m_2} \qquad (4.24)$$

If the coordinate system is rotated through the Euler rotations $\alpha\beta\gamma$ as described in the previous section, each angular momentum eigenfunction transforms under the appropriate rotation matrix according to equation (4.8), and (4.24) becomes

$$\sum_{\mu_1}\sum_{\mu_2} D_{\mu_1m_1}^{j_1} D_{\mu_2m_2}^{j_2} \psi_{j_1\mu_1}\psi_{j_2\mu_2} = \sum_j\sum_\mu C(j_1j_2j; m_1m_2) D_{\mu,m_1+m_2}^j \psi_{j\mu}$$

[10] Wigner, reference 1.

[11] C. Eckart, *Revs. Mod. Phys.* **2**, 305 (1930).

[12] B. L. van der Waerden, *Die Gruppentheoretische Methode in der Quantenmechanik*, Julius Springer, Berlin, 1932.

The common arguments $\alpha\beta\gamma$ of the D-matrices have been dropped for the purposes of brevity. If we use (3.6) for $\psi_{j\mu}$, this equation is now

$$\sum_{\mu_1}\sum_{\mu_2} D^{j_1}_{\mu_1 m_1} D^{j_2}_{\mu_2 m_2}\,\psi_{j_1\mu_1}\,\psi_{j_2\mu_2}$$

$$= \sum_j \sum_\mu \sum_{\mu_1'} C(j_1 j_2 j; m_1 m_2)\, C(j_1 j_2 j; \mu_1'\mu_2')\, D^j_{\mu, m_1+m_2}\psi_{j_1\mu_1'}\psi_{j_2\,\mu-\mu_1'}$$

The sum over μ can be replaced by a sum over $\mu_2' = \mu - \mu_1'$, but then the primes on μ_1' and μ_2' can be dropped since they are only summation indices. Comparison of the coefficients of $\psi_{j_1\mu_1}\,\psi_{j_2\mu_2}$ yields the following result:

$$D^{j_1}_{\mu_1 m_1} D^{j_2}_{\mu_2 m_2} = \sum_j C(j_1 j_2 j; \mu_1\mu_2)\, C(j_1 j_2 j; m_1 m_2)\, D^j_{\mu_1+\mu_2, m_1+m_2} \quad (4.25)$$

This is known as the Clebsch–Gordan series, and is a coupling rule for the D-matrices. The arguments of all D-matrix elements are the same, and the sum is over those values of j for which $\Delta(j_1 j_2 j)$ holds.

The inverse of the Clebsch–Gordan series can be obtained by starting with the inverse relation to (4.24), i.e., equation (3.6), or by direct application of the orthogonality relations for the C-coefficients given in (3.7). Thus, multiplication of (4.25) by $C(j_1 j_2\lambda; \mu_1, \mu-\mu_1)$ and summation over μ_1 keeping the sum $\mu = \mu_1 + \mu_2$ fixed gives

$$\sum_{\mu_1} C(j_1 j_2\lambda; \mu_1, \mu-\mu_1)\, D^{j_1}_{\mu_1 m_1} D^{j_2}_{\mu-\mu_1, m_2}$$

$$= \sum_j \left[\sum_{\mu_1} C(j_1 j_2\lambda; \mu_1, \mu-\mu_1)\, C(j_1 j_2 j; \mu_1, \mu-\mu_1) \right]$$

$$\times C(j_1 j_2 j; m_1 m_2)\, D^j_{\mu, m_1+m_2}$$

or

$$C(j_1 j_2\lambda; m_1 m_2)\, D^\lambda_{\mu, m_1+m_2} = \sum_{\mu_1} C(j_1 j_2\lambda; \mu_1, \mu-\mu_1)\, D^{j_1}_{\mu_1 m_1} D^{j_2}_{\mu-\mu_1, m_2}$$

If now we replace λ by j, multiply by $C(j_1 j_2 j; m_1, m-m_1)$ and sum over m_1, keeping $m = m_1 + m_2$ fixed,

$$\left[\sum_{m'} C(j_1 j_2 j; m_1, m-m_1)\, C(j_1 j_2 j; m_1, m-m_1) \right] D^j_{\mu m}$$

$$= \sum_{\mu_1} \sum_{m_1} C(j_1 j_2 j; \mu_1, \mu-\mu_1)\, C(j_1 j_2 j; m_1, m-m_1)\, D^{j_1}_{\mu_1 m_1} D^{j_2}_{\mu-\mu_1, m-m_1}$$

the inverse of the Clebsch–Gordan series is obtained. This is

$$D^j_{\mu m} = \sum_{\mu_1}\sum_{m_1} C(j_1 j_2 j; \mu_1, \mu-\mu_1)\, C(j_1 j_2 j; m_1, m-m_1)\, D^{j_1}_{\mu_1 m_1} D^{j_2}_{\mu-\mu_1, m-m_1}$$

$$(4.26)$$

This last form permits the in-principle construction of the D-matrix for arbitrary angular momentum j from those for any two angular momenta j_1 and j_2 such that $\Delta(j_1 j_2 j)$ is satisfied. Starting with a knowledge of $D^{\frac{1}{2}}_{m'm}$, all other $D^j_{m'm}$ can be obtained by successive application of the (inverted) Clebsch–Gordan series (4.26). This program is considered more explicitly in section 15.

Another useful point is the connection between the spherical harmonics and the D-matrices for integral angular momentum. First we prove the famous spherical-harmonic addition theorem. Consider the expression

$$g = \sum_m Y^*_{lm}(\theta_1,\varphi_1)\, Y_{lm}(\theta_2,\varphi_2)$$

where (θ_1, φ_1) and (θ_2, φ_2) are the spherical coordinates of two points P_1 and P_2 on the unit sphere. This quantity is invariant under rotations, for, if we rotate the coordinate system through the Euler rotations α, β, γ, we have, from the inverse of equation (2.9),

$$g = \sum_m [R^{-1}\, Y_{lm}(\theta'_1, \varphi'_1)]^*\, R^{-1}\, Y_{lm}(\theta'_2, \varphi'_2)$$

where the primes indicate the spherical coordinates in the rotated frame. Because $D^j_{m'm}(\alpha\beta\gamma)$ is unitary, the definition in equation (4.8) can be rewritten as

$$R^{-1}\psi_{jm} = \sum_{m'} D^{j*}_{mm'}(\alpha\beta\gamma)\, \psi_{jm'}$$

which means that

$$g = \sum_{m_1} \sum_{m_2} \left[\sum_m D^{l*}_{mm_2}\, D^l_{mm_1} \right] Y^*_{lm_1}(\theta'_1, \varphi'_1)\, Y_{lm_2}(\theta'_2, \varphi'_2)$$

But the sum over m is, by the unitary nature of the rotation matrix, simply $\delta_{m_1 m_2}$. Therefore we have shown that g is independent of coordinate system:

$$g = \sum_m Y^*_{lm}(\theta_1, \varphi_1)\, Y_{lm}(\theta_2, \varphi_2) = \sum_m Y^*_{lm}(\theta'_1, \varphi'_1)\, Y_{lm}(\theta'_2, \varphi'_2)$$

i.e., g is invariant under rotations. We can evaluate g in any coordinate system, and the one we now choose is such that P_1 is on the z-axis and $\varphi_2 = 0$; i.e., the xz-plane is defined as the plane containing P_1, P_2 and the origin. Thus $\theta_1 = 0$, and $\theta_2 = \theta$, $\varphi_2 = 0$, θ being the angle between the radius vectors of the two points. Since

$$Y_{lm}(0, \varphi_1) = \delta_{m0} \left(\frac{2l+1}{4\pi} \right)^{\frac{1}{2}},$$

$$g = \left(\frac{2l+1}{4\pi} \right)^{\frac{1}{2}} Y_{l0}(\theta, 0)$$

Finally, the well-known spherical-harmonic addition theorem is obtained

$$Y_{l0}(\theta, 0) = \left(\frac{4\pi}{2l+1}\right)^{\frac{1}{2}} \sum_m Y_{lm}^*(\theta_1, \varphi_1) \, Y_{lm}(\theta_2, \varphi_2) \qquad (4.27)$$

More often we use

$$Y_{l0}(\theta, 0) = \left(\frac{2l+1}{4\pi}\right)^{\frac{1}{2}} P_l(\cos\theta)$$

with P_l the Legendre polynomial, and write the addition theorem as

$$P_l(\cos\theta) = \frac{4\pi}{2l+1} \sum_m Y_{lm}^*(\theta_1, \varphi_1) \, Y_{lm}(\theta_2, \varphi_2) \qquad (4.28)$$

Of course (4.28) applies to any spherical triangle with sides θ, θ_1, θ_2 and dihedral angle $|\varphi_1 - \varphi_2|$ opposite θ.

The connection of all this with the rotation matrices for integral j can now be made directly from their definition in equation (4.8) which, if we consider the point P_2, is

$$Y_{lm'}(\theta_2', \varphi_2') = \sum_m D_{mm'}^l(\alpha\beta\gamma) \, Y_{lm}(\theta_2, \varphi_2) \qquad (4.28a)$$

If the third Euler angle is zero ($\gamma = 0$), then the first two are the spherical coordinates of the new z-axis; i.e., β is the polar angle and α is the azimuthal angle of z' (with respect to the original polar axis). Let us call the intersection of z' with the unit sphere P_1 and thus set $\beta = \theta_1$ and $\alpha = \varphi_1$. Furthermore, if the rotation is carried out so that P_2 is on the x'-axis ($\varphi_2' = 0$), then we may set $m' = 0$ on both sides of the last equation:

$$Y_{l0}(\theta_2', 0) = \sum_m D_{m0}^l(\varphi_1\theta_10) \, Y_{lm}(\theta_2, \varphi_2) \qquad (4.29)$$

This is really the spherical-harmonic addition theorem once more. We obtained this result by considering the spherical harmonic whose arguments are the spherical coordinates of the point P_2 in a rotated coordinate system, the rotation being the one that carries the z-axis into another point P_2. Since $\varphi_2' = 0$, θ_2' is simply the angle θ between P_1 and P_2. Comparison of (4.27) and (4.29) gives the connection

$$D_{m0}^l(\alpha, \beta, 0) = \left(\frac{4\pi}{2l+1}\right)^{\frac{1}{2}} Y_{lm}^*(\beta, \alpha) \qquad (4.30)$$

This result can be used to obtain some very useful properties of the spherical harmonics from those we have already derived for the rotation matrices. For example, application of the symmetry rule in (4.22) shows that

$$Y_{lm}^*(\theta, \varphi) = \left[\left(\frac{2l+1}{4\pi} \right)^{\frac{1}{2}} (-)^m D_{-m,0}^l(\varphi, \theta, 0) \right]^* = [(-)^m Y_{l,-m}^*(\theta, \varphi)]^*$$

and therefore

$$Y_{lm}^*(\theta, \varphi) = (-)^m Y_{l,-m}(\theta, \varphi) \tag{4.31}$$

In Appendix III another proof of this is given. This proof, which involves an explicit closed expression for $Y_{lm}(\theta, \varphi)$, does not depend on the introduction of the rotation matrix, but it is much more lengthy than the one presented here.

We can also use equation (4.30) in the Clebsch–Gordan series (4.25) to obtain the coupling rule for spherical harmonics:

$$Y_{l_1m_1}(\theta, \varphi) \, Y_{l_2m_2}(\theta, \varphi)$$

$$= \sum_l \left[\frac{(2l_1+1)(2l_2+1)}{4\pi(2l+1)} \right]^{\frac{1}{2}} C(l_1l_2l; m_1m_2) \, C(l_1l_2l; 00) \, Y_{l,m_1+m_2}(\theta, \varphi) \tag{4.32}$$

There should be no confusion between this coupling rule and the one in (4.24) with $j_1 = l_1$, $j_2 = l_2$, and $j = l$. In the coupling rule for spherical harmonics in (4.32) the arguments of every spherical harmonic are the same, whereas in equation (4.24) the arguments are different:

$$Y_{l_1m_1}(\theta_1\varphi_1) \, Y_{l_2m_2}(\theta_2\varphi_2) = \sum_l C(l_1l_2l; m_1m_2) \, \psi_{l\,m_1+m_2}(\theta_1, \varphi_1, \theta_2, \varphi_2) \tag{4.33}$$

The ψ_l in (4.33) are wave functions in the composite space $\theta_1\varphi_1$, $\theta_2\varphi_2$ and, of course, are not spherical harmonics. It may be recalled that a basic assumption to the entire discussion in Chapter III of the coupling of two angular momenta is that the two angular momenta commute. This assumption is not fulfilled unless the arguments of the spherical harmonics and the corresponding angular momenta are in different spaces.

The coupling rule for spherical harmonics in (4.32) permits an easy evaluation of the integral of three spherical harmonics.[13] Multiplication of (4.32) by $Y_{l_3m_3}^*(\theta, \varphi)$ and integration over the full solid angle gives

$$\int d\Omega \, Y_{l_3m_3}^* \, Y_{l_2m_2} \, Y_{l_1m_1} = \sum_l \left[\frac{(2l_1+1)(2l_2+1)}{4\pi(2l+1)} \right]^{\frac{1}{2}}$$

$$\times \, C(l_1l_2l; m_1m_2) \, C(l_1l_2l; 00) \int d\Omega \, Y_{l_3m_3}^* \, Y_{l,m_1+m_2} \tag{4.33'}$$

---

[13] Compare, for example, J. A. Gaunt, *Trans. Roy. Soc. London* **A 228,** 195 (1928), where much more laborious procedures were used because explicit forms for the Clebsch–Gordan coefficients were not available.

The spherical harmonics are orthonormal and, therefore, only the $l = l_3$ term of the sum contributes. Consequently,

$$\int d\Omega \, Y^*_{l_3m_3} \, Y_{l_2m_2} \, Y_{l_1m_1}$$

$$= \left[\frac{(2l_1 + 1)(2l_2 + 1)}{4\pi(2l_3 + 1)} \right]^{\frac{1}{2}} C(l_1l_2l_3; m_1m_2m_3) \, C(l_1l_2l_3; 000) \quad (4.34)$$

Again the angular momentum and parity selection rules are operative through the Clebsch–Gordan coefficients $C(l_1l_2l_3; m_1m_2m_3)$ and $C(l_1l_2l_3; 000)$, respectively. An elementary example of the usefulness of this expression is in the matrix elements for dipole radiation between angular momentum states for spinless particles. From equation (III.22) of Appendix III we see that the dipole operator is essentially the spherical harmonic of first order so that the matrix element is proportional to

$$\left[Y_{l_fm_f}, \left(\frac{4\pi}{3}\right)^{\frac{1}{2}} Y_{1m} \, Y_{l_im_i} \right] = \left(\frac{2l_i + 1}{2l_f + 1}\right)^{\frac{1}{2}} C(l_i1l_f; m_imm_f) \, C(l_i1l_f; 000)$$

The subscripts f and i refer, of course, to the final and initial states, respectively. The Clebsch–Gordan coefficients lead to the familiar angular momentum and parity selection rules

$$l_f - l_i = \pm 1 \quad \text{and} \quad m_f - m_i = 0, \pm 1$$

Angular momentum conservation alone would also permit the transition $l_f - l_i = 0$ with $l_i = 0 \to l_f = 0$ absolutely forbidden in a single quantum jump.[14]

15. DETERMINATION OF THE ROTATION MATRICES

We have already pointed out that the inverse of the Clebsch–Gordan series, given in equation (4.26), provides a method for evaluating $D^j_{m'm}$ for arbitrary j once $D^{\frac{1}{2}}_{m'm}$ is known. For instance, we could derive the rotation matrix for $j = \frac{1}{2}$ from the transformation properties of the Pauli spinors and then use equation (4.26) to obtain $D^1_{m'm}$. As a check we can compare this result with that obtained from a study of the transformation properties of the first-order spherical harmonics. Actually, it will be advantageous to carry out this last-mentioned derivation first since

---

[14] Instead of a photon the energy made available in the transition can be used to eject or excite a bound electron. This process is known as internal conversion. For both initial and final states having zero angular momentum only the conversion process can occur. To understand this the finite size of the source (nucleus) of virtual quanta must be taken into account.

its conceptual basis is somewhat simpler. It should be emphasized that the method adopted here is certainly not the simplest procedure for obtaining the elements of D^j for arbitrary j although it is quite simple for the two cases to be treated explicitly: $j = \frac{1}{2}$ and $j = 1$. It does, however, serve to make the significance of the D-matrices clearer. The results for $j = \frac{1}{2}$ are used in Appendix II for the general case.

The rotation matrix with elements $D^1_{m'm}$ can be determined from the transformation properties under rotations of any angular momentum eigenfunction for $j = 1$. Rather than consider the eigenfunctions for an intrinsic spin of one, we use the spherical harmonics Y_{1m} of order one, which are the eigenfunctions of the orbital angular momentum for $l = 1$. The reason for this choice is that the Y_{1m} ($m = 1, 0, -1$) give the position of a point P on the unit sphere, and the transformation properties of the position vector of a point in space are familiar. According to equation (III.22) of Appendix III,

$$Y_{1m}(\theta, \varphi) = \left(\frac{3}{4\pi}\right)^{\frac{1}{2}} \frac{1}{r} \begin{cases} -\dfrac{1}{\sqrt{2}}(x + iy) & m = 1 \\ z & m = 0 \\ \dfrac{1}{\sqrt{2}}(x - iy) & m = -1 \end{cases} \qquad (4.35)$$

The factor $(3/4\pi)^{\frac{1}{2}}1/r$ is invariant under rotations so that the transformation properties of $Y_{1m}(\theta, \varphi)$ are directly determined by those of the indicated combinations of Cartesian coordinates x, y, z of the point P whose spherical coordinates are r, θ, φ. If the coordinate system is rotated through the Euler angles α, β, γ, the Cartesian coordinates x, y, z of P will be changed to x', y', z'. Let the two sets of Cartesian coordinates be written as column matrices $\mathbf{r}$ and $\mathbf{r}'$:

$$\mathbf{r} = \begin{bmatrix} x \\ y \\ z \end{bmatrix}, \qquad \mathbf{r}' = \begin{bmatrix} x' \\ y' \\ z' \end{bmatrix} \qquad (4.36)$$

and let the matrix $M(\alpha\beta\gamma)$ be the 3-by-3 matrix which connects them. Then

$$\mathbf{r}' = M\mathbf{r} \qquad (4.37)$$

In component form this is

$$r'_i = \sum_j M_{ij}r_j$$

Following (4.35), we define the spherical basis in which the vector **V** has the three components V_m $(m = 1, 0, -1)$:

$$\mathbf{V} = \begin{bmatrix} V_1 \\ V_0 \\ V_{-1} \end{bmatrix}, \qquad V_{\pm 1} = \mp \frac{1}{\sqrt{2}}(V_x \pm iV_y), \qquad V_0 = V_z \quad (4.38)$$

There will be a 3-by-3 matrix U which connects the Cartesian and spherical bases,

$$\begin{bmatrix} V_1 \\ V_0 \\ V_{-1} \end{bmatrix} = U \begin{bmatrix} V_x \\ V_y \\ V_z \end{bmatrix}$$

In particular, the vector to the point P has these components in the spherical basis:

$$\mathbf{r}_s = \begin{bmatrix} -\dfrac{1}{\sqrt{2}}(x + iy) \\ z \\ \dfrac{1}{\sqrt{2}}(x - iy) \end{bmatrix}, \qquad \mathbf{r}'_s = \begin{bmatrix} -\dfrac{1}{\sqrt{2}}(x' + iy') \\ z' \\ \dfrac{1}{\sqrt{2}}(x' - iy') \end{bmatrix} \quad (4.39)$$

where obviously the primes refer to the rotated reference frame. The matrix U carries $\mathbf{r}$ into $\mathbf{r}_s$ and $\mathbf{r}'$ into $\mathbf{r}'_s$:

$$\mathbf{r}_s = U\mathbf{r}, \qquad \mathbf{r}'_s = U\mathbf{r}' \quad (4.40)$$

The matrix M_s, by definition, connects the coordinates of P in the rotated spherical basis $(\mathbf{r}'_s)$ with the coordinates in the original spherical basis $(\mathbf{r}_s)$.

$$\mathbf{r}'_s = M_s \mathbf{r}_s$$

From (4.40) and (4.37), we find that

$$\mathbf{r}'_s = U\mathbf{r}' = UM\mathbf{r} = UMU^{-1}\mathbf{r}_s$$

In other words, the unitary transformation U from the Cartesian to the spherical basis transforms the rotation operator M in the usual way

$$M_s = UMU^{-1} \quad (4.41)$$

If we multiply $\mathbf{r}_s$ and $\mathbf{r}'$ by the invariant $(3/4\pi)^{\frac{1}{2}}1/r$, we can then write the transformation properties of the first-order spherical harmonics

$$Y_{1m}(\theta', \varphi') = \sum_{m'} (M_s)_{mm'} Y_{1m'}(\theta, \varphi)$$

But, from the definition of $D^j_{m'm}$ in equation (4.8) and the definition of the rotated function in equation (2.9), we also know that

$$Y_{1m}(\theta', \varphi') = \sum_{m'} D^1_{m'm} Y_{1m'}(\theta, \varphi)$$

Comparison of these two results leads to the conclusion that the rotation matrix for $j = 1$ is the transpose of M_s. In terms of the elements this means that

$$D^1_{m'm} = (M_s)_{mm'} \tag{4.42}$$

To find M_s we first write $M(\alpha\beta\gamma)$ as the product of three Euler rotations

$$M(\alpha\beta\gamma) = M(\gamma)\, M(\beta)\, M(\alpha)$$

where α is a rotation about the original z-axis, β is about the new y-axis, and γ is about the final z-axis. After the first rotation about the z-axis the Cartesian coordinates of P are $x_\alpha, y_\alpha, z_\alpha$:

$$x_\alpha = x \cos \alpha + y \sin \alpha$$

$$y_\alpha = -x \sin \alpha + y \cos \alpha$$

$$z_\alpha = z$$

This transformation is written as a 3-by-3 matrix

$$M(\alpha) = \begin{bmatrix} \cos \alpha & \sin \alpha & 0 \\ -\sin \alpha & \cos \alpha & 0 \\ 0 & 0 & 1 \end{bmatrix}$$

Similarly the matrices $M(\beta)$ and $M(\gamma)$ are

$$M(\beta) = \begin{bmatrix} \cos \beta & 0 & -\sin \beta \\ 0 & 1 & 0 \\ \sin \beta & 0 & \cos \beta \end{bmatrix} ; \qquad M(\gamma) = \begin{bmatrix} \cos \gamma & \sin \gamma & 0 \\ -\sin \gamma & \cos \gamma & 0 \\ 0 & 0 & 1 \end{bmatrix}$$

The product of these three, $M(\alpha\beta\gamma) = M(\gamma)\, M(\beta)\, M(\alpha)$, is

$$M = \begin{bmatrix} \cos \alpha \cos \beta \cos \gamma - \sin \alpha \sin \gamma & \sin \alpha \cos \beta \cos \gamma + \cos \alpha \sin \gamma & -\sin \beta \cos \gamma \\ -\cos \alpha \cos \beta \sin \gamma - \sin \alpha \cos \gamma & -\sin \alpha \cos \beta \sin \gamma + \cos \alpha \cos \gamma & \sin \beta \sin \gamma \\ \cos \alpha \sin \beta & \sin \alpha \sin \beta & \cos \beta \end{bmatrix} \tag{4.43}$$

The unitary transformation from Cartesian to spherical basis is likewise accomplished in three steps.

$$U = cba$$

The matrix a is

$$a = \begin{bmatrix} 1 & 0 & 0 \\ 0 & i & 0 \\ 0 & 0 & 1 \end{bmatrix}$$

and it changes y to iy, leaving x and z unchanged. Thus,

$$a \begin{bmatrix} x \\ y \\ z \end{bmatrix} = \begin{bmatrix} x \\ iy \\ z \end{bmatrix}$$

Next, b is

$$b = \frac{1}{\sqrt{2}} \begin{bmatrix} -1 & -1 & 0 \\ 1 & -1 & 0 \\ 0 & 0 & \sqrt{2} \end{bmatrix} = M\left(\alpha = \frac{5\pi}{4}\right)$$

and it gives the desired linear combinations of x and y:

$$b \begin{bmatrix} x \\ iy \\ z \end{bmatrix} = \begin{bmatrix} -\dfrac{1}{\sqrt{2}}(x + iy) \\ \dfrac{1}{\sqrt{2}}(x - iy) \\ z \end{bmatrix}$$

Finally, c is

$$c = \begin{bmatrix} 1 & 0 & 0 \\ 0 & 0 & 1 \\ 0 & 1 & 0 \end{bmatrix}$$

and it interchanges the two bottom elements:

$$c \begin{bmatrix} -\dfrac{1}{\sqrt{2}}(x + iy) \\ \dfrac{1}{\sqrt{2}}(x - iy) \\ z \end{bmatrix} = \begin{bmatrix} -\dfrac{1}{\sqrt{2}}(x + iy) \\ z \\ \dfrac{1}{\sqrt{2}}(x - iy) \end{bmatrix} = \mathbf{r}_s$$

Multiplying these three together gives $U = cba$:

$$U = \frac{1}{\sqrt{2}} \begin{bmatrix} -1 & -i & 0 \\ 0 & 0 & \sqrt{2} \\ 1 & -i & 0 \end{bmatrix} \tag{4.44}$$

The inverse is $U^{-1} = a^{-1}b^{-1}c^{-1}$, where obviously

$$a^{-1} = \begin{bmatrix} 1 & 0 & 0 \\ 0 & -i & 0 \\ 0 & 0 & 1 \end{bmatrix}, \quad b^{-1} = \frac{1}{\sqrt{2}} \begin{bmatrix} -1 & 1 & 0 \\ -1 & -1 & 0 \\ 0 & 0 & \sqrt{2} \end{bmatrix},$$

$$c^{-1} = \begin{bmatrix} 1 & 0 & 0 \\ 0 & 0 & 1 \\ 0 & 1 & 0 \end{bmatrix}$$

and hence

$$U^{-1} = \frac{1}{\sqrt{2}} \begin{bmatrix} -1 & 0 & 1 \\ i & 0 & i \\ 0 & \sqrt{2} & 0 \end{bmatrix} = U^{+} \tag{4.45}$$

The matrix M_s is now obtained according to (4.41) by direct multiplication of (4.44), (4.43), and (4.45).

$$M_s(\alpha\beta\gamma) = \begin{bmatrix} e^{-i\alpha} \dfrac{1+\cos\beta}{2} e^{-i\gamma} & \dfrac{\sin\beta}{\sqrt{2}} e^{-i\gamma} & e^{i\alpha} \dfrac{1-\cos\beta}{2} e^{-i\gamma} \\[2ex] -e^{-i\alpha} \dfrac{\sin\beta}{\sqrt{2}} & \cos\beta & e^{i\alpha} \dfrac{\sin\beta}{\sqrt{2}} \\[2ex] e^{-i\alpha} \dfrac{1-\cos\beta}{2} e^{i\gamma} & -\dfrac{\sin\beta}{\sqrt{2}} e^{i\gamma} & e^{i\alpha} \dfrac{1+\cos\beta}{2} e^{i\gamma} \end{bmatrix}$$

The rotation matrix $D^1(\alpha\beta\gamma)$ is simply the transpose of this.

$$D^1(\alpha\beta\gamma) = \begin{bmatrix} e^{-i\alpha} \dfrac{1+\cos\beta}{2} e^{-i\gamma} & -e^{-i\alpha} \dfrac{\sin\beta}{\sqrt{2}} & e^{-i\alpha} \dfrac{1-\cos\beta}{2} e^{i\gamma} \\[2ex] \dfrac{\sin\beta}{\sqrt{2}} e^{-i\gamma} & \cos\beta & -\dfrac{\sin\beta}{\sqrt{2}} e^{i\gamma} \\[2ex] e^{i\alpha} \dfrac{1-\cos\beta}{2} e^{-i\gamma} & e^{i\alpha} \dfrac{\sin\beta}{\sqrt{2}} & e^{i\alpha} \dfrac{1+\cos\beta}{2} e^{i\gamma} \end{bmatrix}$$

$$\tag{4.46}$$

By comparison with equation (4.12) it may be seen that the dependence on the angles α and γ is correct. Similarly direct substitution into (4.13) gives complete agreement with (4.46).

We now proceed to evaluate the rotation matrix for $j = \frac{1}{2}$. It is well known that the functions which diagonalize the z-component of the spin operator

$$s_z = \tfrac{1}{2}\sigma_z = \tfrac{1}{2} \begin{bmatrix} 1 & 0 \\ 0 & -1 \end{bmatrix}$$

are the two-component spinors [see remarks following (4.54)]

$$\psi_{\frac{1}{2}} = \begin{pmatrix} 1 \\ 0 \end{pmatrix}, \qquad \psi_{-\frac{1}{2}} = \begin{pmatrix} 0 \\ 1 \end{pmatrix} \tag{4.47}$$

The eigenfunctions which diagonalize some other component are obtained with the rotation matrices. Thus, if χ_m diagonalizes $s_{\mathfrak{f}}$, the component of **s** in the direction [15] $\mathfrak{f}$

$$\chi_m = \sum_{m'} D^{\frac{1}{2}}_{m'm}(\alpha\beta\gamma)\,\psi_{m'}, \quad m = \pm\tfrac{1}{2} \tag{4.48}$$

Here α, β, γ define the Euler rotations which carry the z-axis into the unit vector $\mathfrak{f}$. We shall determine the functions χ_m from the fact that they diagonalize $s_{\mathfrak{f}} = \mathbf{s}\cdot\mathfrak{f}$,

$$(\chi_{m'}, s_{\mathfrak{f}}\chi_m) = m\,\delta_{m'm} \tag{4.49}$$

Because we already know the dependence of $D^{\frac{1}{2}}_{m'm}(\alpha\beta\gamma)$ on α and γ from equation (4.12), we can restrict ourselves to rotations about the y-axis and determine $d^{\frac{1}{2}}_{m'm}(\beta)$.

A good deal of information about $d^{\frac{1}{2}}_{m'm}(\beta)$ can be found just from its unitary character and the fact that its determinant must be $+1$ since it describes a pure rotation with no reflection. If we call the matrix elements A, B, C, D as follows

$$d^{\frac{1}{2}}(\beta) = \begin{pmatrix} A & B \\ C & D \end{pmatrix}$$

then the unitary properties

$$|A|^2 + |B|^2 = 1$$

$$|A|^2 = |D|^2 \quad \text{and} \quad |B|^2 = |C|^2$$

require that $d^{\frac{1}{2}}(\beta)$ have the form

$$d^{\frac{1}{2}}(\beta) = \begin{pmatrix} a\,e^{i\xi} & (1-a^2)^{\frac{1}{2}}\,e^{i\eta} \\ (1-a^2)^{\frac{1}{2}}\,e^{i\zeta} & a\,e^{i\lambda} \end{pmatrix}$$

The quantity a is a non-negative, real number less than or equal to 1. Another unitary property is

$$A^*C + B^*D = 0$$

or

$$a(1-a^2)^{\frac{1}{2}}[e^{i(\zeta-\xi)} + e^{i(\lambda-\eta)}] = 0$$

Since $a = 0$ or $a = 1$ would not give a general unitary matrix this means that

$$e^{i(\lambda+\xi)} = -e^{i(\zeta+\eta)}$$

As a result the determinantal condition

$$AD - BC = 1$$

---

[15] Here, and in other sections as well, we use boldface German letters to indicate unit vectors.

becomes

$$1 = a^2 e^{i(\lambda+\xi)} - (1 - a^2) e^{i(\zeta+\eta)} = e^{i(\lambda+\xi)}$$

Therefore

$$e^{i\lambda} = e^{-i\xi} \quad \text{and} \quad e^{i\xi} = -e^{-i\eta}$$

Because $a \leq 1$, we can introduce another angle ω such that

$$\cos \omega = a \quad \text{and} \quad \sin \omega = (1 - a^2)^{\frac{1}{2}}$$

In this way the most general form of a 2-by-2 unitary matrix whose determinant is $+1$ is

$$d^{\frac{1}{2}}(\beta) = \begin{pmatrix} e^{i\xi} \cos \omega & e^{i\eta} \sin \omega \\ -e^{-i\eta} \sin \omega & e^{-i\xi} \cos \omega \end{pmatrix} \tag{4.50}$$

From (4.47), (4.48), and (4.50) the functions χ_m are

$$\chi_{\frac{1}{2}} = \begin{pmatrix} e^{i\xi} \cos \omega \\ -e^{-i\eta} \sin \omega \end{pmatrix}, \qquad \chi_{-\frac{1}{2}} = \begin{pmatrix} e^{i\eta} \sin \omega \\ e^{-i\xi} \cos \omega \end{pmatrix} \tag{4.51}$$

We are considering a rotation β about the y-axis which carries the z-axis into the direction $\mathfrak{k}$. This is pictured in Fig. 5. The components of $\mathfrak{k}$ are $\mathfrak{k}_x = \sin \beta$, $\mathfrak{k}_y = 0$, $\mathfrak{k}_z = \cos \beta$. The spin operator $\mathbf{s}$ is

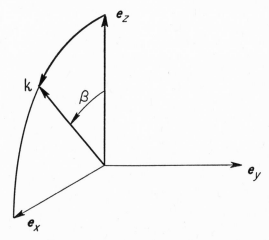

Fig. 5.   Relation of the quantization axes $\mathbf{e}_z$ and $\mathfrak{k}$

$$\mathbf{s} = \frac{1}{2} \begin{pmatrix} \mathbf{e}_z & \mathbf{e}_x - i\mathbf{e}_y \\ \mathbf{e}_x + i\mathbf{e}_y & -\mathbf{e}_z \end{pmatrix}$$

so that

$$s_{\mathfrak{k}} = \mathbf{s} \cdot \mathfrak{k} = \frac{1}{2} \begin{pmatrix} \cos \beta & \sin \beta \\ \sin \beta & -\cos \beta \end{pmatrix} \tag{4.52}$$

This is the matrix representation of $s_\mathfrak{f}$ in the ψ basis. In the (rotated) representation of the χ_m, however, $s_\mathfrak{f}$ is diagonal. That is, the χ_m are the eigenfunctions of $s_\mathfrak{f}$ with the eigenvalues $m = \pm\frac{1}{2}$. For example,

$$s_\mathfrak{f}\chi_\frac{1}{2} = \frac{1}{2}\begin{pmatrix} e^{i\xi}\cos\omega\cos\beta - e^{-i\eta}\sin\omega\sin\beta \\ e^{i\xi}\cos\omega\sin\beta + e^{-i\eta}\sin\omega\cos\beta \end{pmatrix} = \frac{1}{2}\begin{pmatrix} e^{i\xi}\cos\omega \\ -e^{-i\eta}\sin\omega \end{pmatrix}$$

or

$$e^{i\xi}\cos\omega\cos\beta - e^{-i\eta}\sin\omega\sin\beta = e^{i\xi}\cos\omega$$

$$e^{i\xi}\cos\omega\sin\beta + e^{-i\eta}\sin\omega\cos\beta = -e^{-i\eta}\sin\omega$$

The other relation $s_\mathfrak{f}\chi_{-\frac{1}{2}} = -\frac{1}{2}\chi_{-\frac{1}{2}}$ yields the same equations, which are now rewritten as

$$-e^{-i\eta}\sin\omega\sin\beta = e^{i\xi}\cos\omega\,(1 - \cos\beta)$$

$$e^{i\xi}\cos\omega\sin\beta = -e^{-i\eta}\sin\omega\,(1 + \cos\beta)$$

Either one of these yields

$$\tan\omega = -e^{i(\xi+\eta)}\tan\frac{\beta}{2}$$

For the imaginary part of the right side to vanish we must have $\sin(\xi + \eta) = 0$ or $\xi = -\eta + r\pi$, where r is an integer or zero. Therefore

$$\tan\omega = (-)^{r+1}\tan\frac{\beta}{2} = \tan\left[(-)^{r+1}\frac{\beta}{2}\right]$$

and

$$\omega = (-)^{r+1}\frac{\beta}{2} + n\pi$$

In addition to this double multiplicity of solutions, the one remaining phase ξ is not determined. Since its value does not affect the unitary nature of $d_{m'm}^\frac{1}{2}(\beta)$ and the orthonormality of the χ_m, we set $\xi = 0$ (more precisely, $\exp i\xi = 1$), which follows the usual choice of having $d_{m'm}^\frac{1}{2}(\beta)$ and χ_m real. Furthermore we choose $n = 0$ since nothing is gained by greater generality. The χ_m are independent of r. This makes (4.50) agree with equations (4.13) and (4.14).

To summarize, we have shown that the eigenfunctions of

$$s_\mathfrak{f} = \frac{1}{2}\begin{pmatrix} \cos\beta & \sin\beta \\ \sin\beta & -\cos\beta \end{pmatrix} \qquad (4.52)$$

are

$$\chi_{\frac{1}{2}} = \begin{bmatrix} \cos \dfrac{\beta}{2} \\[2ex] \sin \dfrac{\beta}{2} \end{bmatrix}, \qquad \chi_{-\frac{1}{2}} = \begin{bmatrix} -\sin \dfrac{\beta}{2} \\[2ex] \cos \dfrac{\beta}{2} \end{bmatrix} \qquad (4.53)$$

$s_{\mathbf{f}}$ is the component of the spin $\mathbf{s}$ in the direction $\mathbf{f}$, which is in the xz-plane and makes an angle β with the z-axis.   In addition, the rotation matrix for rotations about the y-axis is

$$d^{\frac{1}{2}}(\beta) = \begin{bmatrix} \cos \dfrac{\beta}{2} & -\sin \dfrac{\beta}{2} \\[2ex] \sin \dfrac{\beta}{2} & \cos \dfrac{\beta}{2} \end{bmatrix} \qquad (4.54)$$

The eigenfunctions in (4.53) are called spinors and depend on the half-angle $\beta/2$.   The rotation matrix $d^{\frac{1}{2}}_{m'm}(\beta)$ also depends on $\beta/2$, so that a rotation of 4π, $d^{\frac{1}{2}}_{m'm}(4\pi)$ is the identity matrix, is needed to rotate a spinor into itself.   On the other hand, a rotation of 2π is sufficient to transform a vector or a spherical harmonic into itself.   For this reason, spinors are sometimes referred to as "half-vectors."   From the inverse Clebsch–Gordan series (4.50) it is seen that for all half-integer j there is a double-valuedness in the sense that

$$D^j_{mm'}(\alpha, \beta+2\pi, \gamma) = -D^j_{mm'}(\alpha\beta\gamma)$$

Of course, D^j has the periodicity 4π in β.   For integer j the periodicity in β is 2π.   That is, in general

$$D^j_{mm'}(\alpha, \beta+2\pi, \gamma) = (-)^{2j}D^j_{mm'}(\alpha\beta\gamma)$$

in agreement with (4.13).   Incidentally, it is an almost trivial fact that for all j the periodicity in α and γ is the same as in β; see Appendix II.

The symmetry relations of $d^{\frac{1}{2}}_{m'm}(\beta)$ can be derived directly from (4.54).   Thus,

$$d^{\frac{1}{2}}_{m'm}(\beta) = d^{\frac{1}{2}}_{mm'}(-\beta) = (-)^{m'-m} d^{\frac{1}{2}}_{mm'}(\beta) = (-)^{m'-m} d^{\frac{1}{2}}_{-m',-m}(\beta)$$

Finally, we can derive $D^1_{m'm}(\alpha\beta\gamma)$ from

$$D^{\frac{1}{2}}_{m'm}(\alpha\beta\gamma) = e^{-im'\alpha} d^{\frac{1}{2}}_{m'm}(\beta) e^{-im\gamma}$$

using the inverse of the Clebsch–Gordan series (4.26).   Once again, we need not be concerned with the dependence on α and γ, for, if we substitute equation (4.12) into equation (4.26), we get

$$e^{-i\mu\alpha}\, d_{\mu m}^{j}(\beta)\, e^{-im\gamma} = \sum_{\mu_1}\sum_{m_1} C(j_1 j_2 j; \mu_1, \mu-\mu_1)\, C(j_1 j_2 j; m_1, m-m_1)$$

$$\times\, e^{-i\mu_1\alpha}\, d_{\mu_1 m_1}^{j_1}(\beta)\, e^{-im_1\gamma}\, e^{-i(\mu-\mu_1)\alpha}\, d_{\mu-\mu_1,m-m_1}^{j_2}(\beta)\, e^{-i(m-m_1)\gamma}$$

and the exponentials in α and γ cancel.

For $j_1 = j_2 = \frac{1}{2}$ and $j = 1$ this last result gives

$$d_{\mu m}^{1}(\beta) = \sum_{\mu_1 m_1} C(\tfrac{1}{2}\tfrac{1}{2}1; \mu_1, \mu-\mu_1)\, C(\tfrac{1}{2}\tfrac{1}{2}1, m_1, m-m_1)\, d_{\mu_1 m_1}^{\frac{1}{2}}(\beta)\, d_{\mu-\mu_1,m-m_1}^{\frac{1}{2}}(\beta)$$

$$(4.55)$$

The matrix elements in (4.54) are

$$d_{\frac{1}{2}\frac{1}{2}}^{\frac{1}{2}} = d_{-\frac{1}{2}-\frac{1}{2}}^{\frac{1}{2}} = \cos\frac{\beta}{2}\,; \qquad d_{\frac{1}{2}-\frac{1}{2}}^{\frac{1}{2}} = -d_{-\frac{1}{2}\frac{1}{2}}^{\frac{1}{2}} = -\sin\frac{\beta}{2}$$

and, from Table I.1 of Appendix I, the Clebsch–Gordan coefficients are

$$C(\tfrac{1}{2}\tfrac{1}{2}1; m\pm\tfrac{1}{2}, \mp\tfrac{1}{2}) = \left(\frac{1\mp m}{2}\right)^{\frac{1}{2}}$$

Direct substitution into (4.55) then yields

$$d^1(\beta) = \begin{bmatrix} \dfrac{1+\cos\beta}{2} & -\dfrac{\sin\beta}{\sqrt{2}} & \dfrac{1-\cos\beta}{2} \\[2ex] \dfrac{\sin\beta}{\sqrt{2}} & \cos\beta & -\dfrac{\sin\beta}{\sqrt{2}} \\[2ex] \dfrac{1-\cos\beta}{2} & \dfrac{\sin\beta}{\sqrt{2}} & \dfrac{1+\cos\beta}{2} \end{bmatrix}$$

which agrees with (4.46).

If desired, this procedure can be extended step by step to obtain $d_{m'm}^{j}(\beta)$ for arbitrary j. Of course, this would not be the most advantageous procedure in practice. However, this principle of using a recurrence relation is all that is needed to define all the rotation matrices, and in an explicit manner if need be.

The symmetry properties for $d_{m'm}^{\frac{1}{2}}(\beta)$, which can be derived by inspection of (4.54), can be shown to hold for arbitrary j from (4.55). If, for example, the symmetry relation in equation (4.22) is known to be true for j_1 and j_2, then using it in the right side of equation (4.26) gives

$$\sum_{\mu_1}\sum_{m_1} C(j_1 j_2 j; \mu_1, \mu-\mu_1)\, C(j_1 j_2 j; m_1, m-m_1)\, (-)^{\mu_1-m_1}\, D_{-\mu_1,-m_1}^{j_1*}$$

$$\times (-)^{\mu-\mu_1-m+m_1}\, D_{\mu_1-\mu,m_1-m}^{j_2*} = (-)^{\mu-m}\sum_{\mu_1}\sum_{m_1} C(j_1 j_2 j; -\mu_1, \mu+\mu_1)$$

$$\times\, C(j_1 j_2 j; -m_1, m+m_1)\, D_{\mu_1 m_1}^{j_1*}\, D_{-\mu-\mu_1,-m-m_1}^{j_2*}$$

The second line was obtained by changing the sign of the summation indices. Next we change the signs of the magnetic quantum numbers on the C-coefficients

$$D^j_{\mu m} = (-)^{\mu-m} \sum_{\mu_1} \sum_{m_1} C(j_1 j_2 j; \mu_1, -\mu-\mu_1)$$

$$\times C(j_1 j_2 j; m_1, -m-m_1) D^{j_1*}_{\mu_1 m_1} D^{j_2*}_{-\mu-\mu_1, -m-m_1}$$

There is no phase change because this last operation applied to each C-coefficient gives the phase $(-)^{j_1+j_2-j}$ and the square of this is 1. Now this last equation is just the complex conjugate of equation (4.26) for $-\mu$ and $-m$, which proves our statement that

$$D^{j*}_{\mu m} = (-)^{\mu-m} D^j_{-\mu, -m} \tag{4.22}$$

This relation is now established without recourse to the explicit form given in (4.13).

16. ORTHOGONALITY AND NORMALIZATION OF THE ROTATION MATRICES

The rotation matrices possess the usual orthonormality properties

$$\sum_m D^{j*}_{m'm}(\alpha\beta\gamma) D^j_{m''m}(\alpha\beta\gamma) = \delta_{m'm''}$$

$$\sum_m D^{j*}_{mm'}(\alpha\beta\gamma) D^j_{mm''}(\alpha\beta\gamma) = \delta_{m'm''}$$

These are an expression of the fact that the $D^j_{m'm}(\alpha\beta\gamma)$ are the matrix elements of a unitary transformation from one set of $2j+1$ orthonormal eigenfunctions ψ_{jm} to another set, the rotated eigenfunctions $R(\alpha\beta\gamma)\psi_{jm}$.

Here we show that $D^j_{m'm}(\alpha\beta\gamma)$ are orthogonal on the surface of the unit sphere. Therefore we consider the following integral of two rotation matrices

$$\int d\Omega \, D^{j_1*}_{\mu_1 m_1}(\alpha\beta\gamma) D^{j_2}_{\mu_2 m_2}(\alpha\beta\gamma) \equiv K(j_1 j_2; \mu_1 \mu_2, m_1 m_2) \tag{4.56}$$

The symbol $\int d\Omega$ stands for integration over the full range of the three Euler angles; that is,

$$\int d\Omega = \int_0^{2\pi} d\alpha \int_0^\pi d\beta \sin\beta \int_0^{2\pi} d\gamma$$

The use in (4.56) of the symmetry relation (4.22)

$$D^{j_1*}_{\mu_1 m_1}(\alpha\beta\gamma) = (-)^{\mu_1-m_1} D^{j_1}_{-\mu_1, -m_1}$$

and the Clebsch–Gordan series of (4.25) gives another expression for the integral K:

$$K(j_1 j_2; \mu_1 \mu_2, m_1 m_2) = \sum_j (-)^{\mu_1 - m_1} C(j_1 j_2 j; -\mu_1 \mu_2)$$

$$\times C(j_1 j_2 j; -m_1 m_2) \int d\Omega \, D^j_{-\mu_1 + \mu_2, -m_1 + m_2}(\alpha\beta\gamma) \quad (4.57)$$

The integrations over α and γ can be done immediately since the simple dependence on these angles is known from equation (4.12). Equations (4.56) and (4.57) now become

$$K(j_1 j_2; \mu_1 \mu_2, m_1 m_2) = (2\pi)^2 \, \delta_{\mu_1 \mu_2} \, \delta_{m_1 m_2} \, k(j_1 j_2; \mu_1 \mu_2, m_1 m_2) \quad (4.58a)$$

or

$$K(j_2 j_2; \mu_1 \mu_2, m_1 m_2) = (2\pi)^2 \, \delta_{\mu_1 \mu_2} \, \delta_{m_1 m_2} \sum_j (-)^{\mu_1 - m_1}$$

$$\times C(j_1 j_2 j; -\mu_1 \mu_1) \, C(j_1 j_2 j; -m_1 m_1) \int_0^\pi d\beta \, \sin \beta \, d^j_{00}(\beta) \quad (4.58b)$$

where

$$k(j_1 j_2; \mu_1 \mu_2, m_1 m_2) = \int_0^\pi d\beta \, \sin \beta \, d^{j_1}_{\mu_1 \mu_2}(\beta) \, d^{j_2}_{m_1 m_2}(\beta) \quad (4.58c)$$

From (4.30) we have that

$$d^l_{00}(\beta) = D^l_{00}(0, \beta, 0) = \left(\frac{4\pi}{2l + 1} \right)^{\frac{1}{2}} Y_{l0}(\beta, 0)$$

and, therefore, $d^l_{00}(\beta) = P_l(\cos \beta)$. The integral in (4.58b) is, by the orthogonality of the Legendre polynomials,[16] $2\delta_{j0}$. Comparison of (4.58a) and (4.58b) gives finally

$$k(j_1 j_2; \mu_1 \mu_1, m_1 m_2) = 2(-)^{\mu_1 - m_1} C(j_1 j_2 0; -\mu_1 \mu_1) \, C(j_1 j_2 0; -m_1 m_1)$$

Using the symmetry relation (3.16c) for the C-coefficients, we obtain

$$k(j_1 j_2; \mu_1 \mu_1, m_1 m_1) = (-)^{\mu_1 - m_1} (-)^{j_1 + m_1} (-)^{j_1 + \mu_1} \frac{2}{2j_2 + 1}$$

$$\times C(j_1 0 j_2; -\mu_1, 0, -\mu_1) \, C(j_1 0 j_2; -m_1, 0, -m_1)$$

and, with (3.23), we get

$$k(j_1 j_2; \mu_1 \mu_1, m_1 m_1) = \frac{2}{2j_2 + 1} \, \delta_{j_1 j_2} \quad (4.59)$$

The phase is 1 since $(-)^{\mu_1 - m_1} = (-)^{-\mu_1 + m_1}$ and $(-)^{2(j_1 + m_1)} = 1$.

---

[16] Cf. for example, E. Jahnke and F. Emde, *Tables of Functions*, p. 116, Dover Press, New York, 1943.

Finally, by substituting the result (4.59) into (4.58a), the orthogonality property of the rotation matrices on the unit sphere is expressed in the form

$$\int d\Omega\, D^{j_1*}_{\mu_1 m_1}(\alpha\beta\gamma)\, D^{j_2}_{\mu_2 m_2}(\alpha\beta\gamma) = \frac{8\pi^2}{2j_1 + 1}\, \delta_{\mu_1\mu_2}\, \delta_{m_1 m_2}\, \delta_{j_1 j_2} \qquad (4.60)$$

The factor $8\pi^2$ is the volume of the region of integration and the factor $2j_1 + 1$ is the dimensionality of the set of eigenfunctions $\psi_{j_1 m_1}$. Through (4.30) this reduces to the usual orthonormality of the spherical harmonics:

$$\frac{1}{2\pi}\int d\Omega\, D^{l*}_{m0}(\alpha\beta0)\, D^{l'}_{m'0}(\alpha\beta0) = \frac{4\pi}{2l+1}\, \delta_{mm'}\, \delta_{ll'}$$

and, changing β to θ and α to φ, this is

$$\int_0^{2\pi} d\varphi \int_0^{\pi} d\theta \sin\theta\, Y^*_{l'm'}(\theta, \varphi)\, Y_{lm}(\theta, \varphi) = \delta_{mm'}\, \delta_{ll'} \qquad (4.61)$$

With the aid of the Clebsch–Gordan series we can also evaluate the integral of three rotation matrices in the following manner.

$$\int d\Omega\, D^{j_3*}_{\mu_3 m_3}(\alpha\beta\gamma)\, D^{j_2}_{\mu_2 m_2}(\alpha\beta\gamma)\, D^{j_1}_{\mu_1 m_1}(\alpha\beta\gamma)$$

$$= \sum_j C(j_1 j_2 j; \mu_1\mu_2)\, C(j_1 j_2 j; m_1 m_2) \int d\Omega\, D^{j_3*}_{\mu_3 m_3}(\alpha\beta\gamma)\, D^{j}_{\mu_1+\mu_2, m_1+m_2}(\alpha\beta\gamma)$$

$$= \sum_j C(j_1 j_2 j; \mu_1\mu_2)\, C(j_1 j_2 j; m_1 m_2) \frac{2}{2j_3 + 1}\, \delta_{\mu_1+\mu_2, \mu_3}\, \delta_{m_1+m_2, m_3}\, \delta_{jj_3}$$

Therefore,

$$\int d\Omega\, D^{j_3*}_{\mu_3 m_3}\, D^{j_2}_{\mu_2 m_2}\, D^{j_1}_{\mu_1 m_1}$$

$$= \frac{2}{2j_3 + 1}\, \delta_{\mu_1+\mu_2, \mu_3}\, \delta_{m_1+m_2, m_3}\, C(j_1 j_2 j_3; \mu_1\mu_2)\, C(j_1 j_2 j_3; m_1 m_2) \qquad (4.62)$$

A special case of this is the integral of three spherical harmonics. This can be obtained from (4.62) by using (4.30), and the result checks that given in (4.34).

It is readily seen that the functions $D^j_{mm'}(\alpha\beta\gamma)$ form a complete set in the space of the Euler angles. This is true for each value of j. Consequently these functions can be used as a basis and other quantities expanded in terms of them. Equations (4.8) and (4.28a) can be regarded as examples of such expansions.

V. IRREDUCIBLE TENSORS

17. DEFINITION OF IRREDUCIBLE TENSOR OPERATORS

The subject of irreducible tensor operators occupies a central position in the modern theory of angular momentum. Its importance was first emphasized by Racah who, in three classic papers,[1] derived the algebra of these operators and applied them to studies of atomic spectroscopy. Pioneering work had also been done by Wigner.[2] A tensor is defined by its transformation properties under changes of the coordinate system, and in discussing eigenfunctions of angular momentum and parity, these changes are rotations and inversions. The more familiar Cartesian tensors are not suitable because they usually appear in reducible form. Thus, from direct products of Cartesian tensors, one can form sets of linear combinations of the components of a Cartesian tensor which transform differently. For example, we can form from the nine components T_{ij} of a Cartesian tensor of second rank a scalar, in fact the trace of the tensor,

$$T = \sum_i T_{ii}$$

having one component, an antisymmetric tensor

$$A_k = \tfrac{1}{2}(T_{ij} - T_{ji}) \qquad i, j, k \text{ cyclic}$$

having three components, and a symmetric second-rank tensor with zero trace,

$$S_{ij} = \tfrac{1}{2}(T_{ij} + T_{ji} - \tfrac{2}{3}T \, \delta_{ij})$$

having five independent components. In fact,

$$T_{ij} = \tfrac{1}{3}T \, \delta_{ij} + A_k + S_{ij}$$

The components of the three quantities T, $\mathbf{A}$, and S transform in the same way as the spherical harmonics of order zero, one, and two, respec-

---

[1] G. Racah, *Phys. Rev.* **61**, 186 (1942); **62**, 438 (1942); **63**, 367 (1943).

[2] E. P. Wigner, On the Matrices Which Reduce the Kronecker Products of Representations of Simply Reducible Groups (unpublished); E. P. Wigner, *Am. J. Math.* **63**, 57 (1941).

tively. However, except for rank 0, this representation of the irreducible tensors is not very convenient. The components do not correspond to definite projection quantum numbers. As a very simple case x, y, and z are the components of an irreducible first-rank tensor, but the transformation law is much simpler for the tensors with components $-2^{-\frac{1}{2}}(x + iy)$, z, $2^{-\frac{1}{2}}(x - iy)$ since these are proportional to Y_{1m} with $m = 1$, 0, and -1, respectively. For this reason it will be more useful to discuss irreducible tensors in the spherical coordinate representation, rather than in the Cartesian form. Tensors expressed in this way have been called spherical tensors [3] and some examples of their physical application are given elsewhere.[4]

An irreducible tensor operator of rank L will be defined as a set of $2L+1$ functions T_{LM} ($M = -L, -L+1, \cdots, L$) which transform under the $2L+1$-dimensional representation of the rotation group:

$$R \, T_{LM} R^{-1} = \sum_{M'} D^L_{M'M}(\alpha\beta\gamma) \, T_{LM'} \tag{5.1}$$

The operator $R = \exp(-i\theta \mathbf{n} \cdot \mathbf{J})$ is the rotation operator introduced in section 13, and transforms a wave function ψ into $R\psi$ and an operator Ω into $R\Omega R^{-1}$. The Euler angles which describe the rotation are α, β, γ, and the $D^L_{M'M}(\alpha\beta\gamma)$ are the matrix elements of R in the LM representation. Irreducible tensors as interaction operators, occur in physical application only for integral rank. If they did occur for half-integer rank, transitions would take place between systems obeying Einstein–Bose and Fermi–Dirac statistics.[5]

The definition (5.1) of an irreducible tensor of rank L is simply that it must transform like the spherical harmonic of order L. In this way the three components, V_x, V_y, V_z of a vector form a first rank tensor since the components in the spherical basis, equation (4.38), transform like the spherical harmonics of order one. This follows from the fact that, by definition, the components of a vector transform like the coordinates of a point in space, which in the spherical basis are essentially the

---

[3] M. E. Rose, *Proc. Phys. Soc. London* **A 67**, 239 (1954).

[4] M. E. Rose, *Multipole Fields*, John Wiley & Sons, New York, 1955.

[5] The distinction between wave functions and operators may be somewhat artificial. Thus, in the matrix element $(\psi_{jm}, T_{LM}\psi_{j'm'})$ where T_{LM} commutes with the wave functions, we could equally well discuss the transformation properties of ψ_{jm}. This is clearly an irreducible tensor of rank j, and there is no need to make the restriction to integral j. The major distinction between the integer and half-integer rank tensors is that the latter are double-valued. An obvious example of an irreducible tensor of integer rank L is the spherical harmonic of order L. A case in which the spherical harmonics are involved as tensor operators (as well as wave functions) is in the electrostatic interaction between two electrons, say.

first-order spherical harmonics. A detailed account of first-rank tensors is given in Chapter III of Condon and Shortley.[6] Racah [1] generalizes this algebra to irreducible tensors of arbitrary rank.

The algebra of the spherical tensors in (5.1) has certain analogies to the algebra of Cartesian tensors. Suppose we have a Cartesian tensor $T_{ijk}\cdots$ defined by the transformation law

$$T'_{ijk}\cdots = \sum_{mn\cdots} a_{il}a_{jm}a_{kn} \cdots T_{lmn}\cdots$$

where a_{ij} is the ijth element of an orthogonal 3-by-3 matrix and the prime denotes the components in the rotated reference frame. The rank of the tensor is the number of indices on $T_{ijk}\cdots$ (we make no distinction between covariant and contravariant tensors). Then the sum of two tensors of a given rank is another tensor of the same rank. For instance, for third-rank tensors

$$T_{ijk} + U_{ijk} = V_{ijk}$$

A tensor can be written as the sum of tensors which are symmetric and antisymmetric in a pair of indices:

$$T_{ijk} = \tfrac{1}{2}(T_{ijk} + T_{jik}) + \tfrac{1}{2}(T_{ijk} - T_{jik})$$

The product of two tensors is a tensor whose rank is the sum of the ranks:

$$W_{ijklmn} = T_{ijk} U_{lmn}$$

Finally, the rank of a tensor may be reduced by an even number by making pairs of indices the same and then summing over them. Thus, a third-rank tensor can be contracted into a first-rank tensor, a typical component of which would be

$$\sum_j T_{ijj} \quad \text{or} \quad \sum_j T_{jij} \quad \text{or} \quad \sum_j T_{jji}$$

We now consider the addition and multiplication of the spherical tensors defined in (5.1). Let $T_{L_1M_1}(\mathbf{A}_1)$ and $T_{L_2M_2}(\mathbf{A}_2)$ be two such tensors of rank L_1 and L_2, respectively. The symbols $\mathbf{A}_1$ and $\mathbf{A}_2$ are all the other variables on which the tensors depend. For spherical harmonics, for example, $\mathbf{A}_1$ and $\mathbf{A}_2$ are angular coordinates of two points in space. The addition of spherical tensors is the same as for Cartesian tensors: The sum of two tensors of rank L, $T_{LM}(\mathbf{A}_1) + T_{LM}(\mathbf{A}_2)$ is another tensor of rank L. This follows from the linear nature of the transformation in (5.1).

[6] E. U. Condon and G. H. Shortley, *Theory of Atomic Spectra*, Cambridge University Press, 1935.

Multiplication and contraction, on the other hand, are somewhat different for spherical tensors. The rule is that a tensor of rank L can be constructed from two tensors of rank L_1 and L_2 provided that $\Delta(L_1 L_2 L)$ pertains and that the projection quantum numbers add algebraically.

$$T_{LM}(\mathbf{A}_1, \mathbf{A}_2) = \sum_{M_1} C(L_1 L_2 L; M_1, M - M_1) \, T_{L_1 M_1}(\mathbf{A}_1) \, T_{L_2, M - M_1}(\mathbf{A}_2)$$

$$(5.2)$$

In multiplying spherical tensors, the ranks are thus added vectorially rather than algebraically as is done with Cartesian vectors. Starting with a tensor of first rank, we can also construct a tensor of arbitrary rank by using (5.2). We now prove that T_{LM}, as defined by (5.2), is an irreducible tensor (component) of rank L if $T_{L_1 M_1}$ and $T_{L_2 M_2}$ are irreducible tensor components of rank L_1 and L_2, respectively. Rotating the coordinate system and applying (5.1) to the right side of (5.2), we have

$$R \, T_{LM}(\mathbf{A}_1, \mathbf{A}_2) \, R^{-1} = \sum_{M_1'} \sum_{M_2'} T_{L_1 M_1'}(\mathbf{A}_1) \, T_{L_2 M_2'}(\mathbf{A}_2)$$

$$\times \sum_{M_1} C(L_1 L_2 L; M_1, M - M_1) D_{M_1' M_1}^{L_1} D_{M_2', M - M_1}^{L_2}$$

Next we replace the product of the two D's by the Clebsch–Gordan series of equation (4.25):

$$R \, T_{LM}(\mathbf{A}_1, \mathbf{A}_2) \, R^{-1}$$

$$= \sum_{M_1'} \sum_{M_2'} T_{L_1 M_1'}(\mathbf{A}_1) \, T_{L_2 M_2'}(\mathbf{A}_2) \sum_{M_1} C(L_1 L_2 L; M_1, M - M_1)$$

$$\times \sum_{L'} C(L_1 L_2 L'; M_1' M_2') \, C(L_1 L_2 L'; M_1, M - M_1) \, D_{M_1' + M_2', M}^{L'}$$

$$= \sum_{M_1'} \sum_{M_2'} T_{L_1 M_1'}(\mathbf{A}_1) \, T_{L_2 M_2'}(\mathbf{A}_2) \sum_{L'} C(L_1 L_2 L'; M_1' M_2') \, D_{M_1' + M_2', M}^{L'}$$

$$\times \sum_{M_1} C(L_1 L_2 L; M_1, M - M_1) \, C(L_1 L_2 L'; M_1, M - M_1)$$

The sum over M_1 is $\delta_{LL'}$ by the orthonormality of the C-coefficients. Then

$$R \, T_{LM}(\mathbf{A}_1, \mathbf{A}_2) \, R^{-1} = \sum_{M_1'} \sum_{M_2'} D_{M_1' + M_2', M}^{L}$$

$$\times C(L_1 L_2 L; M_1' M_2') \, T_{L_1 M_1'}(\mathbf{A}_1) \, T_{L_2 M_2'}(\mathbf{A}_2)$$

Letting $M' = M'_1 + M'_2$, we eliminate M'_2,

$$R\,T_{LM}(\mathbf{A}_1, \mathbf{A}_2)\,R^{-1} = \sum_{M'} D^L_{M'M}(\alpha\beta\gamma)$$

$$\times \sum_{M'_1} C(L_1 L_2 L; M'_1, M' - M'_1)\,T_{L_1 M'_1}(\mathbf{A}_1)\,T_{L_2, M'-M'_1}(\mathbf{A}_2)$$

and then the sum over M'_1 (with M' fixed) is $T_{LM'}(\mathbf{A}_1, \mathbf{A}_2)$, according to (5.2). That is

$$R\,T_{LM}(\mathbf{A}_1, \mathbf{A}_2)\,R^{-1} = \sum_{M'} D^L_{M'M}(\alpha\beta\gamma)\,T_{LM'}(\mathbf{A}_1, \mathbf{A}_2)$$

This completes the proof of (5.2). The sum in (5.2) fulfills the definition (5.1) of an irreducible tensor of rank L.

The multiplication and contraction of two spherical tensors of ranks L_1 and L_2 are carried out according to (5.2). The rank of the resulting tensor ranges in integral (usually unit) steps from $L = |L_1 - L_2|$ to $L = L_1 + L_2$. If $L_1 = L_2$, it is possible to construct an invariant, i.e., a tensor of zero rank. From (5.2) with $L = 0$, $L_1 = L_2$, we have

$$T_{00}(\mathbf{A}_1, \mathbf{A}_2) = \sum_{M_1} C(L_1 L_1 0; M_1 - M_1)\,T_{L_1 M_1}(\mathbf{A}_1)\,T_{L_1, -M_1}(\mathbf{A}_2)$$

From equations (3.16c) and (3.23) we see that the C-coefficient is $(-)^{L_1 - M_1}(2L_1+1)^{-\frac{1}{2}}$; the factor $(-)^{L_1}(2L_1+1)^{-\frac{1}{2}}$ can be removed from the sum and absorbed into $T_{00}(\mathbf{A}_1, \mathbf{A}_2)$ to form an invariant $\mathcal{I}_{L_1}$. Dropping the subscripts, we have that

$$\mathcal{I}_L = \sum_M (-)^M\,T_{LM}(\mathbf{A})\,T_{L,-M}(\mathbf{B}) \tag{5.3}$$

is invariant. This invariant was obtained by contracting two tensors of the same rank according to the multiplication law (5.2), and is often called the scalar product of the tensors $T_L(\mathbf{A})$ and $T_L(\mathbf{B})$. For $L = 1$ it is the usual scalar product of two vectors. It is analogous to contraction of Cartesian tensors into invariants according to the following construction:

$$\mathcal{I} = \sum_{ijk} T^*_{ijk}\dots\,U_{ijk}\dots$$

where T and U are Cartesian components of tensors of the same rank. All invariants of this type can be expressed in terms of invariants of the type (5.3).

Perhaps the best known illustration of (5.3) is the spherical harmonic addition theorem, wherein the tensors in (5.3) are spherical harmonics. That $T_{LM} = Y_{LM}$ fulfills (5.1) is obvious from the definition of the latter as the eigenfunction for the orbital angular momentum. From equation (4.28) we see that the invariant in (5.3) is $g_L = (4\pi)^{-1}(2L+1)\,P_L(\cos\theta)$, where θ is the angle between vectors to the points A and B. If we consider the familiar Rayleigh expansion of a plane wave

$$e^{i\mathbf{k}\cdot\mathbf{r}} = \sum_{l=0}^{\infty} i^l(2l+1)\,j_l(kr)\,P_l(\cos\theta)$$

we see that it must be an invariant since $\exp(i\mathbf{k}\cdot\mathbf{r})$ is a function of the scalar product of two vectors. This invariance can be also exhibited as the contraction of spherical tensors by using the spherical harmonic addition theorem in the right-hand side.

$$e^{i\mathbf{k}\cdot\mathbf{r}} = 4\pi \sum_{l=0} i^l j_l(kr) \sum_{m} (-)^m\, Y_{lm}(\theta_{\mathbf{r}},\varphi_{\mathbf{r}})\, Y_{l-m}(\theta_{\mathbf{k}},\varphi_{\mathbf{k}}) \qquad (5.4)$$

The arguments of the spherical harmonics, $(\theta_{\mathbf{r}}, \varphi_{\mathbf{r}})$ and $(\theta_{\mathbf{k}}, \varphi_{\mathbf{k}})$, are the angular coordinates of the position and wave vectors, respectively.

It is important to distinguish between zero-rank tensors like (5.3) which are invariant in form and scalars in the sense in which this term is used here. The zero-rank tensor is a combination of operators $T_0(\mathbf{A}_1\mathbf{A}_2 \cdots)$ so constructed that it is equal to $T_0(\mathbf{A}_1'\mathbf{A}_2' \cdots)$, where $\mathbf{A}_1'$ is the form of the operator $\mathbf{A}_i$ in a new reference frame obtained from the original one by a rotation. This is in contrast to a scalar (field) where, as discussed in section 4 for example, the functional form is changed while the numerical value is fixed. For example, a scalar field which is expressible as $Y_1^0(\cos\theta)$ clearly vanishes on the equatorial plane $(\theta = \pi/2)$ defined relative to a given polar axis; but the same field would be expressed differently (as a linear combination of Y_{1m}) if a different polar axis were used. It is important to bear in mind that a scalar, as the term is used here, may be a tensor of any rank.

It is also of interest to make a distinction between a vector and its components. Thus, if x_k represent the components of the vector $\mathbf{r}$ while $\mathbf{e}_k$ are the unit vectors along the axes, then

$$\mathbf{r} = \sum_k \mathbf{e}_k x_k = \sum_k \mathbf{e}_k' x_k'$$

where the primes refer to a rotated system. Hence the vector $\mathbf{r}$ is an invariant and a tensor of zero rank. Consequently, a distinction is also to be made between tensors of rank one (the x_k, for example) and the vector which is a contraction in the sense of (5.3) of the components x_k and the unit vectors $\mathbf{e}_k$. In general, the components of a vector may be tensors of any rank and this situation will be pertinent when multipole fields are discussed; see section 27.

18. RACAH'S DEFINITION OF IRREDUCIBLE TENSOR OPERATORS

In the previous section we defined irreducible tensors by their transformation properties under rotations. Racah [1] has given the following definition in terms of commutation rules of the tensor components with the angular momentum operators. The set T_{LM} constitutes an irreducible tensor if the commutator relationships

$$[J_x \pm iJ_y, T_{LM}] = [(L \mp M)(L \pm M + 1)]^{\frac{1}{2}} T_{LM \pm 1} \qquad (5.5a)$$

$$[J_z, T_{LM}] = M T_{LM} \qquad (5.5b)$$

are fulfilled. We shall now show that this definition is equivalent to the one already given in (5.1).

Consider first an infinitesimal rotation $\delta\phi$ about the z-axis. According to our previous definition (5.1), the rotated tensor components are

$$R \, T_{LM} \, R^{-1} = e^{-i\delta\phi J_z} \, T_{LM} \, e^{i\delta\phi J_z} = \sum_{M'} D^L_{M'M}(\delta\phi, 0, 0) \, T_{LM'} \qquad (5.6)$$

On the right side $D^L_{M'M}(\delta\phi, 0, 0) = \exp(-iM \, \delta\phi) \, \delta_{M'M}$, since a rotation about the quantization introduces only a phase change. Expanding the exponentials in

$$e^{-i\delta\phi J_z} \, T_{LM} \, e^{i\delta\phi J_z} = e^{-iM\delta\phi} \, T_{LM}$$

to first order in $\delta\phi$ gives

$$(1 - i \, \delta\phi \, J_z) \, T_{LM} (1 + i \, \delta\phi \, J_z) = (1 - iM \, \delta\phi) \, T_{LM}$$

$$T_{LM} - i \, \delta\phi \, J_z \, T_{LM} + i \, \delta\phi \, T_{LM} \, J_z = T_{LM} - iM \, \delta\phi \, T_{LM}$$

This is just the second commutation rule above in (5.5b):

$$[J_z, T_{LM}] = M T_{LM}$$

The more complicated rule with $J_\pm = J_x \pm iJ_y$ is obtained by performing infinitesimal rotations about x and y. An infinitesimal rotation of $\delta\theta$ about the x-axis is accomplished by the Euler rotations $\alpha = -\pi/2$, $\beta = \delta\theta$, $\gamma = \pi/2$:

$$e^{-i\delta\theta J_x} \, T_{LM} \, e^{i\delta\theta J_x} = \sum_{M'} D^L_{M'M}\left(-\frac{\pi}{2}, \delta\theta, +\frac{\pi}{2}\right) T_{LM'}$$

From equations (4.11) and (4.12) we have that

$$D^L_{M'M}\left(-\frac{\pi}{2}, \delta\theta, \frac{\pi}{2}\right) = e^{iM'(\pi/2)}(LM'|e^{-i\delta\theta J_y}|LM) \, e^{-iM(\pi/2)}$$

Substituting this and expanding up to first order in $\delta\theta$ gives

$$T_{LM} - i\,\delta\theta[J_x,\,T_{LM}]$$

$$= \sum_{M'} e^{i(M'-M)\,(\pi/2)}(LM'|\,1 - i\,\delta\theta\,J_y\,|LM)\,T_{LM'}$$

$$= \sum_{M'} e^{i(M'-M)\,(\pi/2)}[\delta_{M'M} - i\,\delta\theta(LM'|\,J_y\,|LM)]\,T_{LM'}$$

Canceling the leading terms on each side gives, finally

$$[J_x,\,T_{LM}] = \sum_{M'} e^{i(M'-M)\,(\pi/2)}(LM'|\,J_y\,|LM)\,T_{LM'} \tag{5.7}$$

An infinitesimal rotation $\delta\theta$ about y simply gives

$$e^{-i\delta\theta J_y}\,T_{LM}\,e^{i\delta\theta J_y} = \sum_{M'} D^L_{M'M}(0,\,\delta\theta,\,0)\,T_{LM'}$$

$$= \sum_{M'} (LM'|\,e^{-i\delta\theta J_y}\,|LM)\,T_{LM'}$$

which implies that

$$[J_y,\,T_{LM}] = \sum_{M'} (LM'|\,J_y\,|LM)\,T_{LM'} \tag{5.8}$$

We have only to substitute into (5.5) and (5.7) the matrix elements of J_y in the LM representation, equation (2.28).

$$[J_x,\,T_{LM}] = \tfrac{1}{2}\,\{[(L-M)(L+M+1)]^{\frac{1}{2}}\,T_{L,M+1}$$

$$+ [(L+M)(L-M+1)]^{\frac{1}{2}}\,T_{L,M-1}\}$$

$$[J_y,\,T_{LM}] = \frac{1}{2i}\,\{[(L-M)(L+M+1)]^{\frac{1}{2}}\,T_{L,M+1}$$

$$- [(L+M)(L-M+1)]^{\frac{1}{2}}\,T_{L,M-1}\}$$

The commutation rules for the linear combinations $J_x \pm iJ_y$ are now

$$[J_x + iJ_y,\,T_{LM}] = [(L-M)(L+M+1)]^{\frac{1}{2}}\,T_{L,M+1}$$

$$[J_x - iJ_y,\,T_{LM}] = [(L+M)(L-M+1)]^{\frac{1}{2}}\,T_{L,M-1}$$

These agree with (5.5a) above. We have therefore proved that the definition of irreducible tensor operators by their transformation properties under rotations implies the commutation properties with the angular momentum operators given in (5.5), which is Racah's definition.

A familiar example of the commutation relations (5.5) is that of the angular momentum operators themselves. In the spherical basis introduced in equation (4.38) of section 15 the three components J_M ($M = 1$, 0, -1) of the angular momentum are

$$J_1 = -\frac{1}{\sqrt{2}}(J_x + iJ_y) = -\frac{1}{\sqrt{2}}J_+$$

$$J_0 = J_z \tag{5.9}$$

$$J_{-1} = \frac{1}{\sqrt{2}}(J_x - iJ_y) = \frac{1}{\sqrt{2}}J_-$$

The commutation rules for the angular momentum operators are given in equations (2.19) and (2.24):

$$[\mathbf{J}^2, J_M] = 0$$

$$[J_z, J_\pm] = \pm J_\pm \qquad [J_+, J_-] = 2J_z$$

These can be rewritten as

$$[J_x + iJ_y, J_1] = 0 \qquad\qquad [J_x - iJ_y, J_1] = \sqrt{2}\,J_0$$

$$[J_x + iJ_y, J_0] = \sqrt{2}\,J_1 \qquad [J_x - iJ_y, J_0] = \sqrt{2}\,J_{-1} \tag{5.10a}$$

$$[J_x + iJ_y, J_{-1}] = \sqrt{2}\,J_0 \qquad [J_x - iJ_y, J_{-1}] = 0$$

and

$$[J_z, J_{\pm 1}] = \pm J_{\pm 1} \qquad [J_z, J_0] = 0 \tag{5.10b}$$

Comparison with equations (5.5a) and (5.5b) show that (5.10a) and (5.10b) are just the commutation relations for a first-rank tensor. Similarly, the relations $[J_M, \mathbf{J}^2] = 0$ imply that $\mathbf{J}^2$ is a scalar. Note that $\mathbf{J}^2 = \sum_M (-)^M J_M J_{-M}$, and compare with (5.3).

It will prove useful later on to write the commutation rules (5.5) completely in the spherical basis. Thus we multiply (5.5a) by $\mp 1/\sqrt{2}$ to get

$$[J_{\pm 1}, T_{LM}] = \mp [\tfrac{1}{2}(L \mp M)(L \pm M + 1)]^{\frac{1}{2}} T_{L,\,M\pm 1}$$

$$[J_0, T_{LM}] = M\, T_{LM}$$

Referring to the table of Clebsch–Gordan coefficients in Appendix I, we see that

$$C(L1L; M+\mu, -\mu)$$

$$= \frac{1}{[L(L+1)]^{\frac{1}{2}}} \begin{cases} [\tfrac{1}{2}(L-M)(L+M+1)]^{\frac{1}{2}} & \mu = 1 \\ M & \mu = 0 \\ -[\tfrac{1}{2}(L+M)(L-M+1)]^{\frac{1}{2}} & \mu = -1 \end{cases} \tag{5.11}$$

This permits the two equations, (5.5a) and (5.5b), to be written as

$$[J_\mu, T_{LM}] = (-)^\mu C(L1L; M+\mu, -\mu)[L(L+1)]^{\frac{1}{2}} T_{L, M+\mu} \quad (5.12)$$

In a similar way the matrix elements of the angular moment operators, equation (2.28), can be written in spherical basis as

$$(j'm'|J_\mu|jm) = \delta_{j'j}\, \delta_{m',m+\mu}(-)^\mu[j(j+1)]^{\frac{1}{2}} C(j1j; m+\mu, -\mu) \quad (5.13)$$

This is a general rule; in section 20 it will be useful for the purpose of identifying the intrinsic spin eigenfunctions for $s = 1$.

19. THE WIGNER–ECKART THEOREM

Having defined the components T_{LM} of the irreducible tensor in (5.1) and derived their commutation relations with the angular momentum operators in (5.5), we shall next discuss the matrix elements of such tensor components between states of sharp angular momentum. The matrix elements of irreducible tensors are of great importance, since they occur in all processes wherein states of sharp angular momentum are involved. The Wigner–Eckart theorem to be discussed in this section is of key significance for several reasons. First, as was mentioned before, it brings into explicit form the dependence of all matrix elements of irreducible tensors on the projection quantum numbers. Therefore it permits a separation of those features of a physical process that depend on the geometry or symmetry properties of the system from those depending on the detailed physical description. Second, the theorem constitutes a formal expression of the important conservation laws of angular momentum. The manner in which the formalism of angular momentum theory expresses this conservation rule is via the dependence of the matrix element on the projection quantum numbers. The Wigner–Eckart theorem [7] states that the dependence of the matrix element $(j'm'|T_{LM}|jm)$ on the projection quantum numbers is entirely contained in the Clebsch–Gordan coefficient:

$$(j'm'|T_{LM}|jm) = C(jLj'; mMm')(j' \| T_L \| j) \quad (5.14)$$

The quantity $(j' \| T_L \| j)$ is called the reduced matrix element of the set of tensor operators T_{LM}. As the notation implies, it is independent of M, m, and m'. The conservation of angular momentum is contained in the Clebsch–Gordan coefficient, which vanishes unless

$$\Delta(j\, L\, j') \quad \text{and} \quad m' = M + m$$

[7] E. P. Wigner, *Gruppentheorie*, Friedrich Vieweg und Sohn, Braunschweig, 1931. C. Eckart, *Revs. Mod. Phys.* **2**, 305 (1930).

Since $j'm'$ play the role of the quantum numbers of the final state and jm those of the initial state, the interpretation to be given to L and M, in accordance with the conservation laws, is the set of quantum numbers for the absorbed radiation. This, in fact, is just the interpretation which, the detailed applications show, is the correct one. When the radiation is emitted the only change to be made is that $-M$ is the projection quantum number in place of M. This is in accord with the result that absorption is replaced by emission if T_{LM} is replaced by $(T_{LM})^+$, the Hermitian adjoint of T_{LM}, and the latter quantity will be seen to be proportional to $T_{L,-M}$ for all Hermitian operators. Investigation of this point may be left as an exercise for the reader.

The proof of the Wigner–Eckart theorem rests on the commutation relations for the components T_{LM} of an irreducible tensor with the angular momentum operators, which have been given in (5.5). The matrix elements of the second commutation relation, (5.5b), are

$$(j'm'|J_z T_{LM}|jm) - (j'm'|T_{LM} J_z|jm) = M(j'm'|T_{LM}|jm)$$

In the first matrix element let J_z operate on the left (i.e., on $\psi_{j'm'}$), and in the second let J_z operate on the right (on ψ_{jm}); then, transposing the right-hand side,

$$(m' - m - M)(j'm'|T_{LM}|jm) = 0 \tag{5.15}$$

Therefore $(j'm'|T_{LM}|jm) = 0$ unless $m' = M + m$. This is the conservation rule for projection quantum numbers.

The matrix elements of the first commutator (5.5a) are

$$(j'm'|J_\pm T_{LM}|jm) - (j'm'|T_{LM} J_\pm|jm)$$
$$= [(L \mp M)(L \pm M + 1)]^{\frac{1}{2}} (j'm'|T_{L,M\pm 1}|jm)$$

(The raising and lowering operators $J_\pm$ should be distinguished from the irreducible tensor components $J_{\pm 1}$.) In the first matrix element let $J_\pm$ operate on the left as $J_\mp$, and in the second let $J_\pm$ operate on the right; the matrix elements of $J_\pm$ are given in equation (2.28).

$$[(j' \pm m')(j' \mp m' + 1)]^{\frac{1}{2}} (j'm' \mp 1|T_{LM}|jm)$$
$$- [(j \mp m)(j \pm m + 1)]^{\frac{1}{2}} (j'm'|T_{LM}|jm \pm 1)$$
$$= [(L \mp M)(L \pm M + 1)]^{\frac{1}{2}} (j'm'|T_{L,M\pm 1}|jm) \tag{5.16}$$

According to (5.15), each term in this equation vanishes unless $m' = M + m \pm 1$. To understand this relation, let us consider the coupling of two angular momenta j and L to give a resultant j'. According to the method of introduction of the Clebsch–Gordan coefficients, equation (3.4), the corresponding wave functions are related as follows:

$$\psi_{j'm'} = \sum_m \sum_M C(jLj'; mMm') \psi_{jm} \psi_{LM} \qquad (5.17)$$

The only non-vanishing terms on the right are those that satisfy conservation of angular momentum

$$m' = m + M, \qquad \Delta(jLj')$$

If $J'_{\mp} = J_{\mp} + L_{\mp}$ is applied to both sides of (5.17), and use is made of the matrix elements in equation (2.28), we find that

$$[(j' \pm m')(j' \mp m' + 1)]^{\frac{1}{2}} \psi_{j\,m'\mp1} = \sum_m \sum_M [(j \pm m)(j \mp m + 1)]^{\frac{1}{2}}$$

$$\times\, C(jLj'; mMm') \psi_{j\,m\mp1} \psi_{LM} + \sum_m \sum_M [(L \mp M)(L \pm M + 1)]^{\frac{1}{2}}$$

$$\times\, C(jLj'; mMm') \psi_{jm} \psi_{L,M\mp1}$$

On the left we use (5.17) to eliminate $\psi_{j'\,m'\mp1}$; on the right we replace $m \mp 1$ with μ and M with λ in the first member and replace m with μ and $M \mp 1$ with λ in the second member.

$$\sum_\mu \sum_\lambda [(j' \pm m')(j' \mp m' + 1)]^{\frac{1}{2}} C(jLj'; \mu\lambda, m'\mp1) \psi_{j\mu} \psi_{L\lambda}$$

$$= \sum_\mu \sum_\lambda [(j \mp \mu)(j \pm \mu + 1)]^{\frac{1}{2}} C(jLj'; \mu\pm1, \lambda m') \psi_{j\mu} \psi_{L\lambda}$$

$$+ \sum_\mu \sum_\lambda [(L \mp M)(L \pm M + 1)]^{\frac{1}{2}} C(jLj'; \mu, \lambda\pm1, m') \psi_{j\mu} \psi_{L\lambda}$$

Equating the coefficients on both sides of the $\mu = m$, $\lambda = M$ term, and transposing the first term on the right, we obtain

$$[(j' \pm m')(j' \mp m' + 1)]^{\frac{1}{2}} C(jLj'; mM, m'\mp1)$$

$$- [(j \mp m)(j \pm m + 1)]^{\frac{1}{2}} C(jLj'; m\pm1, Mm')$$

$$= [(L \mp M)(L \pm M + 1)]^{\frac{1}{2}} C(jLj'; m, M\pm1, m') \quad (5.18)$$

This is a recurrence relation for the Clebsch–Gordan coefficients for changes in the projection quantum numbers; the values of j, j', and L, however, do not change.[8] Nothing in the derivation restricts (5.18) to integral values of L. On the other hand, it is necessary that $\Delta(j\,L\,j')$

---

[8] See also Appendix I. It is essential to note that the two equations obtained by writing (5.18) with upper and lower signs are sufficient to determine the dependence of the C-coefficients on the projection quantum numbers.

holds. Comparison of equations (5.18) and (5.16) reveals that the dependence of the projection quantum numbers of the tensor matrix elements and of the C-coefficients is exactly the same in both equations. Also, in both equations all the terms vanish unless $m' = m + M \pm 1$. Therefore the dependence of the matrix elements $(j'm' \,|\, T_{LM} \,|\, jm)$ on the magnetic quantum numbers is given by the Clebsch–Gordan coefficient $C(jLj'; mMm')$:

$$(j'm' \,|\, T_{LM} \,|\, jm) = C(jLj'; mMm')(j' \,\|\, T_L \,\|\, j) \qquad (5.14)$$

To recapitulate, the C-coefficient in (5.14) contains the angular momentum conservation laws, for, according to the rules for the vanishing of C-coefficients,

$$(j'm' \,|\, T_{LM} \,|\, jm) = 0 \quad \text{unless} \quad \Delta(j \, L \, j') \text{ and } m + M = m'$$

For example, the tensor components T_{LM} may be the potentials of a 2^L-pole Maxwell field, as we shall discuss in more detail in Chapter VII. Then L is the angular momentum of the emitted or absorbed radiation.

The factor that distinguishes the matrix elements of one tensor interaction from another of the same rank is the reduced matrix element $(j' \,\|\, T_L \,\|\, j)$, since the geometric factor $C(jLj'; mMm')$ is always the same. The reduced matrix element contains the physical properties of the tensor field and those of the initial and final systems. It is in this way that the separation of the physical and geometrical features of many physical problems is achieved. As a case in point, the shape of the angular distribution of radiation emitted by oriented nuclei or that of the angular correlation of cascade radiation (Chapter X) can be specified completely, without discussion of details of nuclear structure, whenever the radiations emitted can be treated as pure angular momentum waves. When this is not true, there will be more than one reduced matrix element involved in the angular distribution, and the shape of the distribution depends on ratios of these parameters.

In many cases the j and j' dependence of the reduced matrix elements is easily calculated. As an example of a simplified nature we consider a diatomic rotator. From equation (4.34), which gives the matrix elements of the spherical harmonics between eigenstates of the orbital angular momentum, we find

$$(l_f m_f \,|\, Y_{lm} \,|\, l_i m_i) = C(l_i l l_f; m_i m m_f) \, C(l_i l l_f; 000) \left[\frac{(2l_i + 1)(2l + 1)}{4\pi(2l_f + 1)} \right]^{\frac{1}{2}}$$

$$(5.19a)$$

According to the Wigner–Eckart theorem, the reduced matrix elements of the spherical harmonics between states of zero intrinsic spin are given by

$$(l_f \parallel Y_l \parallel l_i) = C(l_i l l_f; 000) \left[\frac{(2l_i + 1)(2l + 1)}{4\pi(2l_f + 1)} \right]^{\frac{1}{2}} \quad (5.19b)$$

For a symmetric top the calculation of the reduced matrix elements of Y_{lm} follows readily from (4.62) and (4.30).

As another interesting example, we consider the angular momentum operators. From (5.13) we can determine the reduced matrix of the angular momentum component J_μ.

$$(j'm' | J_\mu | jm) = \delta_{jj'} (-)^\mu [j(j + 1)]^{\frac{1}{2}} C(j'1j; m', -\mu, m)$$

Use of the symmetry relations equations (3.17a) and (3.16a) for the Clebsch–Gordan coefficients gives (with $j = j'$)

$$C(j'1j; m', -\mu, m) = (-)^{1-\mu}(-)^{j+1-j'} C(j1j'; m\mu m')$$

Therefore

$$(j'm' | J_\mu | jm) = \delta_{jj'}[j(j + 1)]^{\frac{1}{2}} C(j1j'; m\mu m') \quad (5.19c)$$

and the reduced matrix element of the angular momentum is

$$(j' \parallel J \parallel j) = [j(j + 1)]^{\frac{1}{2}} \delta_{jj'} \quad (5.20)$$

We were able to deduce this rather simple result because of the definition of the operators J_μ, and, in this way, the details of constructing specific wave functions are completely avoided.

We use the Wigner–Eckart theorem to obtain some results of interest for the coupling of a charged particle with a radiation field. The results which are directly obtained are the selection rules for emission or absorption of radiation. This depends on an interaction operator which is a zero-rank tensor and must, therefore, be expressible as a sum of invariants, each of which is a contraction of irreducible tensors [3,4] as in (5.3). If this interaction operator is written in the form

$$H' = \sum_L \sum_M (-)^M T_{L,-M}(\mathbf{X}) T_{LM}(\mathbf{Y}) \quad (5.21)$$

the variables $\mathbf{X}$ and $\mathbf{Y}$ referring to the field and source, the terms corresponding to a given value L are associated with radiation of angular momentum L. The notation in (5.21) is not meant to imply that both tensors have identical construction.

The matrix elements of H' between states of specified angular momentum j_f and j_i for such radiation are then

$$(j_f m_f | T_{LM}(\mathbf{Y}) | j_i m_i)$$

and, by the Wigner–Eckart theorem, this yields the triangular condition

$$\Delta(j_f \, L \, j_i) \tag{5.22a}$$

for the angular momentum conservation rule, and the reduced matrix element would contain

$$\pi_f \pi_L \pi_i = 1 \tag{5.22b}$$

for the parity rule. Here π_f is the parity of the final state with angular momentum j_f, π_i is the parity of the initial state and π_L that of the radiation; π_f etc., is $+1$ for even parity and -1 for odd parity.

The operators $T_{LM}(\mathbf{Y})$ constitute the multipole moments of the system and elementary examples will be given below. If the emitted radiation has zero intrinsic spin the operators $T_{LM}(\mathbf{X})$ are simply $Y_{LM}(\mathbf{f})$ where $\mathbf{f}$ is the unit propagation vector of the emitted field. For intrinsic spin unity, one needs to couple the orbital angular momentum eigenfunctions Y_{LM} with those of the intrinsic spin in the usual way. This problem is deferred until after the latter are investigated in section 21. However, to provide some specific examples, we can consider the alternative procedure wherein the Cartesian form of the tensors is employed.

We write for the interaction with electromagnetic radiation

$$H' = -\frac{1}{c} \mathbf{A} \cdot \mathbf{j}$$

where $\mathbf{A}$ is the vector potential and $\mathbf{j}$ the current density operator for the source particles. For propagation along $\mathbf{k}$ we have

$$\mathbf{A} = \boldsymbol{\epsilon} \, e^{i\mathbf{k} \cdot \mathbf{r}}$$

and $\boldsymbol{\epsilon}$ is a polarization vector. We now expand the exponential

$$e^{i\mathbf{k} \cdot \mathbf{r}} = 1 + i\mathbf{k} \cdot \mathbf{r} - \tfrac{1}{2}(\mathbf{k} \cdot \mathbf{r})^2 \cdots$$

and consider each term separately. The first gives a contribution to H' proportional to $\boldsymbol{\epsilon} \cdot \mathbf{j}$ and, since $\mathbf{j}$ is the only factor depending on the source variables and because it is a polar vector, this term is recognized as (part of) the electric dipole contribution.[9] This means $L = 1$ and $\pi_L = -1$. From the next term we get a contribution proportional to

$$\boldsymbol{\epsilon} \cdot \mathbf{j} \, \mathbf{k} \cdot \mathbf{r} = \sum_{st} \epsilon_s k_t j_s r_t$$

the indices s and t serving to distinguish the Cartesian components. The factors $j_s r_t$ are source operators, and the bilinear combination is of even parity. But these nine quantities $j_s r_t$ do not correspond to a single value

---

[9] Other terms of this character would arise from the $(\mathbf{k} \cdot \mathbf{r})^2$ and other even powers in the expansion of the exponential.

of L. Instead, by decomposing this second-rank tensor into its irreducible constituents, as was done above, we recognize the existence of two relevant parts, one with $L = 1$ and one with $L = 2$. For both, $\pi_L = 1$. The first is the magnetic dipole and the second the electric quadrupole contributions.

If we consider the next term, contributions to multipoles with $\pi_L = -1$ and $L = 1$, 2, and 3 arise. These are electric octupole, magnetic quadrupole, and the afore-mentioned correction to the electric dipole.[9] This method of expressing the tensor in irreducible form is rather clumsy and not readily generalized. It is also clear that, unless kr_{ave} is small compared to unity, we run into trouble with the retardation expansion. There are an infinite number of terms to examine, and the expansion is not one like that contemplated in (5.21). In (5.21) only a finite number of terms need to be considered in any case, because fixing j_f and j_i requires L to lie between the limits $|j_f - j_i|$ and $j_f + j_i$. If the retardation expansion is valid, the smallest value of L consistent with parity requirements may and often does suffice. In any event, we rarely need more than two values of L.

Still another and perhaps the most compelling argument against the use of the Cartesian form of the tensors becomes apparent when we attempt to express higher-rank tensors in irreducible form. For example, the problem of expressing the components of the third-rank tensor $a_i b_j c_k$, which is the direct product of three first-rank tensors, in terms of irreducible tensors of rank ≤ 3 is somewhat complicated and does not readily lend itself to simple generalization.[10] The fact that there are only two first-rank tensors, $\mathbf{j}$ and $\mathbf{r}$, in the example discussed here, and that all higher-rank tensors are constructed from these does not materially simplify the problem.

This discussion points up the extreme utility of using tensors in the spherical basis. For the electromagnetic field this will be done in section 27. For spinless particles the Rayleigh expansion of the plane wave is all that is needed. Thus, in

$$e^{i\mathbf{k}\cdot\mathbf{r}} = 4\pi \sum_{lm} i^l j_l(kr)\, Y_{lm}(\hat{\mathbf{f}})\, Y_{lm}^*(\hat{\mathbf{r}}) \qquad (5.23)$$

[10] It should be recognized that in constructing the irreducible tensors in Cartesian form all traces, formed by equating two indices and summing over the common index, must vanish. As will become evident, this is automatically accomplished when the spherical form of the tensors is used. For an example of the use of the higher-order Cartesian tensors see E. Greuling, *Phys. Rev.* **61**, 568 (1942). The importance of the higher-rank tensors in radiative transitions in nuclei is made evident by the existence of isomeric states from which de-excitation occurs only by the emission of radiations of large angular momentum.

the higher terms in the retardation expansion that belong to a given angular momentum are automatically collected in the spherical Bessel function $j_l(kr)$. The corresponding expansion for photons is given in section 27; see equation (7.38). These expansions automatically introduce the spherical tensors as in (5.21), and the selection rules (5.22) follow immediately from the Wigner–Eckart theorem. Obviously no complication arises from treating the general case of arbitrary angular momentum emitted or absorbed.

The static moments can also be discussed in terms of the Wigner–Eckart theorem. We consider the simplest multipole moments of a quantum-mechanical state ψ_{jm}, the electric and magnetic dipole moments, and the electric quadrupole moment.[11] For a single particle, the electric dipole operator is $\mathbf{p} = e\mathbf{r}$, and the magnetic dipole operator is $\boldsymbol{\mu} = \mu_0(g_L\mathbf{L} + g_S\mathbf{S})$, where $\mathbf{L}$ and $\mathbf{S}$ are the orbital and spin angular momentum operators, μ_0 is a magneton unit, and g_L and g_S are the appropriate gyromagnetic ratios. These operators are both first-rank tensors, but there is an important difference between them; the electric dipole operator $\mathbf{p}$ is a polar vector, $\pi_1 = -1$, whereas the magnetic dipole operator is an axial vector or pseudo-vector. It does not change sign under an inversion of the coordinate system, i.e., $\pi_1 = 1$. The difference is therefore a parity property. If ψ_{jm} is a many-particle state, the electric and magnetic dipole operators are simply the vector sum of all the one-particle operators. Thus, $\mathbf{p} = \sum_k \mathbf{p}_k$, where k is a label numbering the particles, is a first-rank tensor and is, moreover, a polar vector; and $\boldsymbol{\mu} = \sum_k \boldsymbol{\mu}_k$ is a first-rank tensor and is, in fact, an axial vector. The expectation values of these operators are simply the matrix elements $(jm\,|\,\mathbf{p}\,|\,jm)$ and $(jm\,|\,\boldsymbol{\mu}\,|\,jm)$, and will vanish unless $\Delta(j\,1\,j)$. This requires $j \geq \frac{1}{2}$. In addition, the state ψ_{jm} is expected to have a definite parity, and therefore the electric dipole moment will vanish since it connects only states of different parity. Therefore, in a state of definite angular momentum j and parity π, we have the following results:

Electric dipole moment, $(\pi jm\,|\,\mathbf{p}\,|\,\pi jm) = 0$
Magnetic dipole moment, $(\pi jm\,|\,\boldsymbol{\mu}\,|\,\pi jm) = 0$ unless $j \geq \frac{1}{2}$, $\pi_1 = 1$

According to the Wigner–Eckart theorem (5.14), the expectation value of the Mth component of $\boldsymbol{\mu}$ in the spherical basis is

$$(jm\,|\,\mu_M\,|\,jm) = C(j1j;\,mMm)(j\,\|\,\mu\,\|\,j)$$

---

[11] The effects of higher moments are usually quite small. However, recently evidence for the nuclear magnetic octupole moment has been presented. See V. Jaccarino, J. G. King, R. A. Satten, and H. H. Stroke, *Phys. Rev.* **94**, 1798 (1954); P. Kusch and T. G. Eck, *Phys. Rev.* **94**, 1799 (1954).

Thus M is necessarily zero from the rules for the non-vanishing of the Clebsch–Gordan coefficients. The only non-vanishing component is $\mu_0 = \mu_z$, the component of $\boldsymbol{\mu}$ on the quantization axis. Using the table of Clebsch–Gordan coefficients from Table I.1, we then have

$$(jm|\mu_z|jm) = m[j(j+1)]^{-\frac{1}{2}} (j \| \mu \| j) \qquad (5.24)$$

For the totality of the $2j+1$ degenerate states there is but one quantity measured. What is called the magnetic moment is the expectation value of μ_z for the maximum projection quantum number [12]

$$\mu = (jj|\mu_z|jj) = [j/(j+1)]^{\frac{1}{2}} (j \| \mu \| j)$$

In an analogous way the electric quadrupole moment of a particle is the second-rank tensor $Q_{2M} = er^2 P_{2M}(\cos\theta)$, which is an even operator. It connects only states of the same parity, $\pi_2 = 1$. Furthermore, summing over many particles does not alter these properties. The matrix elements of the components of the quadrupole tensor vanish unless $\Delta(j\,2\,j)$ and $M = 0$.

Electric quadrupole moment: $(\pi jm|Q_{2M}|\pi jm) = 0$ unless $j \geq 1$

Once more, the Wigner–Eckart theorem (5.14) gives

$$(jm|Q_{2M}|jm) = C(j2j; m0m)(j \| Q_2 \| j)$$

The Clebsch–Gordan coefficient can be found from (3.19) or reference 6.

$$(jm|Q_{2M}|jm) = \frac{3m^2 - j(j+1)}{[j(j+1)(2j-1)(2j+3)]^{\frac{1}{2}}} (j \| Q_2 \| j) \quad (5.25)$$

The measured quadrupole moment is given as the expectation value in the state with maximum projection quantum number

$$Q = \frac{3j^2 - j(j+1)}{[j(j+1)(2j-1)(2j+3)]^{\frac{1}{2}}} (j \| Q_2 \| j)$$

$$= \left[\frac{j(2j-1)}{(j+1)(2j+3)}\right]^{\frac{1}{2}} (j \| Q_2 \| j)$$

---

[12] The matrix element of the magnetic moment in some other state ψ_{jm} is then simply

$$\mu_m = \frac{m}{j}\,\mu$$

Use is made of this result in calculating moments by the "method of diagonal sums"; see, e.g., E. P. Wigner, *The j–j Coupling Shell Model for Nuclei*, University of Wisconsin, 1951.

The quadrupole moment matrix element in another substate is then [11]

$$Q_m = \frac{3m^2 - j(j+1)}{3j^2 - j(j+1)} \, Q = \frac{3m^2 - j(j+1)}{j(2j-1)} \, Q$$

In general, there are only even electric multipole moments and odd magnetic multipole moments. In addition, the matrix elements of the 2^L-pole moments vanish unless $j \geq \frac{1}{2}L$; see section 29.

20. THE PROJECTION THEOREM FOR FIRST-RANK TENSORS

In extension of the principles developed in the preceding sections, we now prove and discuss a useful theorem having to do with the matrix element of an arbitrary first-rank tensor $\mathbf{T}_1$ in the angular momentum representation. The commutation rules of an irreducible tensor of rank L with the angular momentum operators are given in (5.12); for a first-rank tensor we set $L = 1$. The projection of a first-rank tensor on the angular momentum, with the latter also a first-rank tensor, is proportional to the scalar product or contraction of these first-rank tensors,

$$\mathbf{J} \cdot \mathbf{T}_1 = \sum_\mu (-)^\mu J_\mu \, T_{1,-\mu} \tag{5.26}$$

The actual projection, obtained by dividing by the magnitude of the angular momentum, is $[j(j+1)]^{-\frac{1}{2}} (\mathbf{J} \cdot \mathbf{T}_1)$. The theorem we are considering is concerned with this operator and has the following three forms:

I. Decomposition Theorem of the First Kind

$$(j'm' \,|\, T_{1M} \,|\, jm) = \frac{(jm' \,|\, J_M(\mathbf{J} \cdot \mathbf{T}_1) \,|\, jm)}{j(j+1)} \, \delta_{jj'} \tag{5.27}$$

II. Factorization Theorem

$$(j'm' \,|\, J_M(\mathbf{J} \cdot \mathbf{T}_1) \,|\, jm) = (j'm' \,|\, J_M \,|\, jm)(j\|\mathbf{J} \cdot \mathbf{T}_1\|j)\delta_{jj'} \tag{5.28}$$

III. Decomposition Theorem of the Second Kind

$$(jm' \,|\, T_{1M} \,|\, jm) = \frac{(jm' \,|\, J_M \,|\, jm)(j \,\|\, \mathbf{J} \cdot \mathbf{T}_1 \,\|\, j)}{j(j+1)} \tag{5.29}$$

The decomposition theorem I is the primary result to be demonstrated. From it II and III follow readily. The theorem expressed in (5.27) has a rather simple meaning. To see this one can express the tensor $\mathbf{T}_1$ as follows:

$$\mathbf{T}_1 = \mathbf{J}^1(\mathbf{J}^1 \cdot \mathbf{T}_1) + (\mathbf{T}_1)_\perp \tag{5.30}$$

where

$$\mathbf{J}^1 = [j(j+1)]^{-\frac{1}{2}} \, \mathbf{J}$$

is a unit angular momentum operator: $(\mathbf{J}^1)^2$ has the eigenvalue one. The first term on the right-hand side in (5.30) can be interpreted as the component of $\mathbf{T}_1$ along $\mathbf{J}$. The second term is the component perpendicular to $\mathbf{J}$. The result expressed in (5.27) is

$$(j'm'|T_{1M}|jm) = (jm'|J_M^1(\mathbf{J}^1 \cdot \mathbf{T}_1)|jm)\, \delta_{j'j}$$

It therefore states that the component perpendicular to the angular momentum does not contribute to the expectation value of the vector operator between states of the same j. This theorem is basic to the vector model of the atom, where the perpendicular component averages to zero as the result of its precession about the total angular momentum.

To prove the decomposition theorem of the first kind (I), consider the matrix element

$$\mathfrak{M} = (j'm'|J_M(\mathbf{J} \cdot \mathbf{T}_1)|jm) \tag{5.31}$$

Expanding the scalar product of the two tensors according to (5.3)

$$\mathfrak{M} = \sum_\mu (-)^\mu (j'm'|J_M J_\mu T_{1,-\mu}|jm)$$

and introducing the commutator of J_μ and $T_{1,-\mu}$, the matrix element $\mathfrak{M}$ becomes

$$\mathfrak{M} = \sum_\mu (-)^\mu (j'm'|J_M\, T_{1,-\mu}J_\mu|jm) + \sum_\mu (-)^\mu (j'm'|J_M[J_\mu, T_{1,-\mu}]|jm) \tag{5.31a}$$

We first consider the second term which, on use of the commutation relations (5.12), becomes

$$\sqrt{2}\, (j'm'|J_M T_{10}|jm) \sum_\mu C(111; 0, -\mu)$$

It is easy to see that the sum over μ of $C(111; 0, -\mu)$ vanishes. In fact, this sum is a special case of

$$\sum_\mu (-)^\mu\, C(lls; \mu, -\mu) \tag{5.32}$$

because

$$C(111; 0, -\mu) = (-)^{\mu+1}\, C(111; \mu, -\mu)$$

by (3.17a). Using

$$C(ll0; \mu, -\mu) = (-)^{l-\mu}(2l+1)^{-\frac{1}{2}}$$

we see that the sum in (5.32) is

$$(-)^l(2l+1)^{\frac{1}{2}} \sum_\mu C(lls; \mu, -\mu)\, C(ll0; \mu, -\mu)$$

$$= (-)^l(2l+1)^{\frac{1}{2}}\, \delta_{s0} \tag{5.32a}$$

In the present consideration $s = 1$ which establishes the result stated above.

This leads to the conclusion that the second term in (5.31a) vanishes and

$$\mathfrak{M} = \sum_\mu (-)^\mu (j'm' | J_M\, T_{1,-\mu}\, J_\mu | jm)$$

Next we operate with J_μ on ψ_{jm} in the matrix element on the right-hand side and with the adjoint of $J_M(J_M{}^+ = (-)^M J_{-M})$ on $\psi_{j'm'}$. Using the matrix elements given in (5.13), we find

$$\mathfrak{M} = [j(j + 1)\, j'(j' + 1)]^{\frac{1}{2}} C(j'1j'; m'-M, M)$$

$$\times \sum_\mu (j'm' - M | T_{1,-\mu} | jm + \mu)\, C(j1j; m+\mu, -\mu)$$

Application of the Wigner–Eckart theorem, (5.14), yields

$$\mathfrak{M} = [j(j + 1)j'(j' + 1)]^{\frac{1}{2}}(j' \parallel T_1 \parallel j)\, C(j'1j'; m'-M, M)$$

$$\times \sum_\mu C(j1j'; m+\mu, -\mu)\, C(j1j; m+\mu, -\mu)$$

$$= [j(j + 1)j'(j' + 1)]^{\frac{1}{2}}(j' \parallel T_1 \parallel j)\, C(j'1j'; mM)$$

$$\times \sum_\mu C(j1j'; m+\mu, -\mu)\, C(j1j; m+\mu, -\mu)$$

The sum over μ is $\delta_{jj'}$, by the orthonormality of the C-coefficients, so that

$$\mathfrak{M} = \delta_{jj'}\, \delta_{m+M,m'} j(j + 1)\, C(j1j; mM)(j' \parallel T_1 \parallel j)$$

In the remainder of this section the Kronecker delta for the projection quantum numbers will be understood. Another application of the Wigner–Eckart theorem shows that

$$\mathfrak{M} = \delta_{jj'} j(j + 1)(j'm' | T_{1M} | jm)$$

This result relates the matrix elements of $\mathbf{T}_1$ between states of the same total angular momentum with the corresponding matrix elements of $J_M(\mathbf{J \cdot T}_1)$; see (5.31).

Therefore,

$$(jm' | T_{1M} | jm) = \frac{(jm' | J_M(\mathbf{J \cdot T}_1) | jm)}{j(j + 1)} \tag{5.27}$$

This completes the proof of I.

With (5.27) established, the factorization theorem of (5.28) follows readily. In detail

$$(j'm' | J_M(\mathbf{J \cdot T}_1) | jm) = \sum_{j''m''} (j'm' | J_M | j''m'')(j''m'' | \mathbf{J \cdot T}_1 | jm) \tag{5.33}$$

On the right-hand side we use the fact that $\mathbf{J} \cdot \mathbf{T}_1$ is a zero-rank tensor and

$$(j''m''|\mathbf{J}\cdot\mathbf{T}_1|jm) = \delta_{j''j}\, \delta_{m''m}(j \,\|\, \mathbf{J}\cdot\mathbf{T}_1 \,\|\, j)$$

Then

$$(j'm'|J_M(\mathbf{J}\cdot\mathbf{T}_1)|jm) = \delta_{jj'}(jm'|J_M|jm)(j \,\|\, \mathbf{J}\cdot\mathbf{T}_1 \,\|\, j) \qquad (5.28)$$

which is precisely (5.28).

The decomposition theorem III, as written in (5.29), is obtained by substituting (5.28) into (5.27). This is perhaps a somewhat more familiar form of the theorem than that given in (5.27). It states that, aside from a reduced matrix element, the matrix of any first-rank tensor is the same as that of the angular momentum operators. This is easily understood since the components of $\mathbf{J}$ constitute a first-rank tensor, and by the Wigner–Eckart theorem we have the following result for the components of any pair of such tensors, $\mathbf{A}_1$ and $\mathbf{B}_1$,

$$\frac{(j'm'|A_{1M}|jm)}{(j' \,\|\, A_1 \,\|\, j)} = \frac{(j'm'|B_{1M}|jm)}{(j' \,\|\, B_1 \,\|\, j)} \qquad (5.34)$$

The result (5.29) is then a consequence of

$$(j' \,\|\, \mathbf{J}\cdot\mathbf{T}_1 \,\|\, j) = [j(j+1)]^{\frac{1}{2}}(j \,\|\, T_1 \,\|\, j)\, \delta_{jj'}$$

and

$$(j' \,\|\, J \,\|\, j) = [j(j+1)]^{\frac{1}{2}}\, \delta_{jj'}$$

see (5.20).

A typical application of the above considerations occurs in the calculation of expectation value of the magnetic moment in a state of sharp total angular momentum. Let $\mathbf{J}$, $\mathbf{L}$, and $\mathbf{S}$, be the operators for the total, orbital, and intrinsic spin angular momenta, respectively. Our remarks apply to the situation where $\mathbf{J}^2$, J_z, $\mathbf{L}^2$, $\mathbf{S}^2$ are diagonal. Then the total angular momentum is

$$\mathbf{J} = \mathbf{L} + \mathbf{S}$$

and the magnetic moment is

$$\boldsymbol{\mu} = \mu_0(g_L\mathbf{L} + g_S\mathbf{S})$$

where μ_0 is a magneton unit and g_L, g_S are appropriate gyromagnetic ratios. Then the expectation value of the Mth component of $\boldsymbol{\mu}$ ($M = 1, 0, -1$) can be obtained from the decomposition theorem of the second kind,

$$(jm|\mu_M|jm) = \frac{(jm|J_M|jm)(j \,\|\, \boldsymbol{\mu}\cdot\mathbf{J} \,\|\, j)}{j(j+1)} \qquad (5.35)$$

Now

$$\boldsymbol{\mu}\cdot\mathbf{J} = \mu_0(g_L\mathbf{L} + g_S\mathbf{S})\cdot(\mathbf{L} + \mathbf{S}) = \mu_0[g_L\mathbf{L}^2 + g_S\mathbf{S}^2 + (g_L + g_S)\mathbf{L}\cdot\mathbf{S}]$$

and $2\mathbf{L}\cdot\mathbf{S} = \mathbf{J}^2 - \mathbf{L}^2 - \mathbf{S}^2$. So

$$\frac{\mathbf{\mu}\cdot\mathbf{J}}{\mu_0} = \frac{1}{2}[(g_L + g_S)\mathbf{J}^2 + (g_L - g_S)(\mathbf{L}^2 - \mathbf{S}^2)]$$

The matrix elements of $\mathbf{J}^2$ are

$$(j'm'|\mathbf{J}^2|jm) = \delta_{jj'}\,\delta_{mm'}j(j+1)$$

The Wigner–Eckart theorem also gives an expression for the matrix elements of the zero-rank tensor $\mathbf{J}^2$,

$$(j'm'|\mathbf{J}^2|jm) = C(j0j';m0m')(j'\parallel \mathbf{J}^2 \parallel j)$$

$$= \delta_{mm'}\,\delta_{jj'}(j\parallel \mathbf{J}^2 \parallel j)$$

Therefore, $(j\parallel \mathbf{J}^2 \parallel j) = j(j+1)$, which is not very surprising. Finally, the reduced matrix element of $\mathbf{\mu}\cdot\mathbf{J}$ is (in units μ_0)

$$(j\parallel \mathbf{\mu}\cdot\mathbf{J} \parallel j)$$

$$= \tfrac{1}{2}\{(g_L + g_S)\,j(j+1) + (g_L - g_S)[l(l+1) - s(s+1)]\} \quad (5.36)$$

The diagonal matrix elements of J_M are

$$(jm|J_M|jm) = \delta_{M0}(jm|J_0|jm) = m\,\delta_{M0} \quad (5.37)$$

Combining equations (5.35), (5.36), and (5.37), we obtain for the expectation value of μ_M

$$(jm|\mu_M|jm) = m\,\delta_{M0}\left[\frac{1}{2}(g_L + g_S) + \frac{1}{2}(g_L - g_S)\,\frac{l(l+1) - s(s+1)}{j(j+1)}\right]$$

$$(5.38)$$

The quantity in square brackets is the Landé g-factor. The observed magnetic moment is the above matrix element for $m = j$. The quantity to compare with this is

$$\mu = (jj\parallel \mu_{M=0} \parallel jj) = \frac{(j\parallel \mathbf{\mu}\cdot\mathbf{J} \parallel j)}{j+1}$$

where the last equality follows from (5.29) and (5.37). It should be emphasized that the only way in which a physical model enters is in the evaluation of the gyromagnetic ratios g_L and g_S.

21. ANGULAR MOMENTUM OF A VECTOR FIELD

The concept of scalar fields is a fairly familiar one. As an example, consider the function $R(r)\,Y_{lm}$ defined at all points of three-dimensional space or simply Y_{lm} defined at all points on a sphere. For a given l and m only a single function is defined. If we are concerned with transformations such as rotations of the coordinate axes, the $2l+1$ functions for each value of l are involved. In any event, the angular momentum operator for the scalar field is $\mathbf{L} = -i\,\mathbf{r} \times \nabla$.

We are also interested in vector fields. A prime example is the electromagnetic field discussed specifically in Chapter VII. Here at each point of space, or a suitably restricted portion thereof, we define three functions as components of a vector. It will be useful to restrict our attention to the situation in which these three functions transform under D^L: that is, they are each irreducible tensors of rank L. The total set of such tensors in a vector field would number $3(2L+1)$ functions. The interesting fact is that these functions can be regrouped into three new sets of $2J+1$ functions where $J = L + 1$, L, and $L - 1$ (assuming $L > 0$) and that each of these sets constitute the components of an irreducible tensor of rank J. This suggests that the angular momentum for a vector field can be represented by the operator

$$\mathbf{J} = \mathbf{L} + \mathbf{S}$$

where $\mathbf{L} = -i\,\mathbf{r} \times \nabla$ is the orbital angular momentum operator and $\mathbf{S}$ is a set of 3-by-3 matrices representing the intrinsic spin. The functions discussed above are the angular momentum eigenfunctions, and in addition there will be a three-component spinor function which is the eigenfunctions associated with $\mathbf{S}$. Also the eigenvalue of $\mathbf{S}^2$ is $S(S + 1) = 2$ or $S = 1$. These are the results that we wish to obtain in this section.

We now consider the angular momentum of a vector field $\mathbf{V}(\mathbf{r})$, which consists of three functions of the space coordinates. In the Cartesian representation of the vector field, $\mathbf{V}(\mathbf{r})$ is written

$$\mathbf{V}(\mathbf{r}) = \begin{bmatrix} V_1(\mathbf{r}) \\ V_2(\mathbf{r}) \\ V_3(\mathbf{r}) \end{bmatrix}$$

and the indices 1, 2, and 3 replace the usual x, y, and z. In section 4 we discussed a rotation in terms of a change in the orientation of the coordinate system. Below, Fig. 6b represents the effect of a rotation R on Fig. 6a; the only difference in the two figures is in the orientation of the reference axes: the vectors are not altered by the rotation. The position vectors $\mathbf{r}$ and $\mathbf{r}'$ locate the same point in space, and $\mathbf{V}$ and $\mathbf{V}'$ are the same vector as seen in a fixed reference frame. Primes have been added in Fig. 6b, however, because the components with respect to the new coordinate axes are different from the components in the original coordinate system. Thus the very same point in space has the Cartesian coordinates $x_1x_2x_3$ in the original system and $x_1'x_2'x_3'$ in the rotated system. Similarly the components of the vector are $V_1(x_1x_2x_3)$, $V_2(x_1x_2x_3)$, $V_3(x_1x_2x_3)$ and $V_1'(x_1'x_2'x_3')$, $V_2'(x_1'x_2'x_3')$, $V_3'(x_1'x_2'x_3')$ in the two systems.

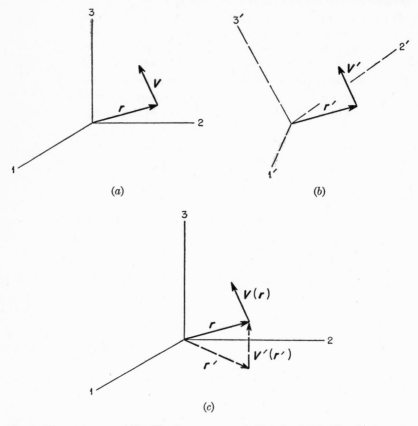

Fig. 6. The vectors **r** and **V** in Fig. 6a are represented by **r'** and **V'** in Fig. 6b where the coordinate system has been rotated from the 1–2–3 axes to the 1'–2'–3' axes. In Fig. 6c the coordinate system is the same as in Fig. 6a, and the vectors are rotated into **r'** and **V'**.

The two sets of components are related by a matrix M which, as an example, for a rotation about x_3, is

$$M = \begin{bmatrix} \cos\theta & \sin\theta & 0 \\ -\sin\theta & \cos\theta & 0 \\ 0 & 0 & 1 \end{bmatrix}$$

We can, on the other hand, consider a rotation in another way. That is, a rotation of the coordinate system, in which the Cartesian components of a point and a vector undergo the transformation

$$\left.\begin{array}{l} \mathbf{r}:\ x_1x_2x_3 \\ \mathbf{V(r)}:\ V_1(x_1x_2x_3),\ V_2(_1xx_2x_3),\ V_3(x_1x_2x_3) \end{array}\right\} \xrightarrow[\rightarrow]{M} \left\{\begin{array}{l} \mathbf{r'}:\ x_1'x_2'x_3' \\ \mathbf{V'(r')}:\ V_1'(x_1'x_2'x_3'),\ V_2'(x_1'x_2'x_3'),\ V_3'(x_1'x_2'x_3') \end{array}\right.$$

is equivalent to a rotation of a vector $\mathbf{r'}$ and a vector $\mathbf{V'}$ keeping the coordinate axes fixed, in which the vectors $\mathbf{r'}$ and $\mathbf{V'(r')}$ are carried into the vectors $\mathbf{r}$ and $\mathbf{V(r)}$, respectively:

$$\left.\begin{array}{l} \mathbf{r'}:\ x_1'x_2'x_3' \\ \mathbf{V'(r')}:V_1'(x_1'x_2'x_3'),\ V_2'(x_1'x_2'x_3'),\ V_3'(x_1'x_2'x_3') \end{array}\right\} \xrightarrow[\rightarrow]{M} \left\{\begin{array}{l} \mathbf{r}:\ x_1x_2x_3 \\ \mathbf{V(r)}:\ V_1(x_1x_2x_3),\ V_2(x_1x_2x_3),\ V_3(x_1x_2x_3) \end{array}\right.$$

Figure 6c illustrates this equivalent rotation. In this way the rotated vector of the rotated coordinates $\mathbf{V(r)}$ is obtained from the original vector of the original coordinates $\mathbf{V'(r')}$ by application of the familiar transformation M, an example of which has been given above.

$$M\ \mathbf{V'(r')}\ =\ \mathbf{V(r)}$$

Similarly, M carries $\mathbf{r'}$ into $\mathbf{r}$: $\mathbf{r} = M\mathbf{r'}$. Therefore

$$\mathbf{V'(r')}\ =\ M^{-1}\ \mathbf{V}(M\mathbf{r'})$$

but, since this equation applies to any point, the prime can be dropped from $\mathbf{r'}$.

$$\mathbf{V'(r)}\ =\ M^{-1}\ \mathbf{V}(M\mathbf{r}) \tag{5.39}$$

Furthermore, $\mathbf{V'(r)} = R\ \mathbf{V(r)}$, where $R = e^{-i\theta(\mathbf{n}\cdot\mathbf{J})}$ is the rotation operator of section 5 and $\mathbf{J}$ is the angular momentum operator for the field. Therefore

$$R\ \mathbf{V(r)}\ =\ M^{-1}\ \mathbf{V}(M\mathbf{r}) \tag{5.40}$$

Equation (5.40) exhibits the manner in which the angular momentum of a vector field, contained in R, may be deduced from the transformation properties of the field under rotations, which are determined by M. In particular, consider an infinitesimal rotation through the angle ϵ about the x_3-axis; then the equations (5.40) become

$$V_1'(x_1'x_2'x_3')\ =\ V_1(x_1{+}x_2\epsilon,\ -x_1\epsilon{+}x_2,\ x_3)\ -\ \epsilon\ V_2(x_1{+}x_2\epsilon,\ -x_1\epsilon{+}x_2,\ x_3)$$

$$V_2'(x_1'x_2'x_3')\ =\ \epsilon\ V_1(x_1{+}x_2\epsilon,\ -x_1\epsilon{+}x_2,\ x_3)\ +\ V_2(x_1{+}x_2\epsilon,\ -x_1\epsilon{+}x_2,\ x_3)$$

$$V_3'(x_1'x_2'x_3')\ =\ V_3(x_1{+}x_2\epsilon,\ -x_1\epsilon{+}x_2,\ x_3)$$

Expanding to first order in ϵ, this becomes

$$V_1'(x_1'x_2'x_3') = V_1(x_1x_2x_3) + \epsilon\left(x_2\frac{\partial}{\partial x_1} - x_1\frac{\partial}{\partial x_2}\right)V_1(x_1x_2x_3) - \epsilon V_2(x_1x_2x_3)$$

$$V_2'(x_1'x_2'x_3') = V_2(x_1x_2x_3) + \epsilon\left(x_2\frac{\partial}{\partial x_1} - x_1\frac{\partial}{\partial x_2}\right)V_2(x_1x_2x_3) + \epsilon V_1(x_1x_2x_3)$$

$$V_3'(x_1'x_2'x_3') = V_3(x_1x_2x_3) + \epsilon\left(x_2\frac{\partial}{\partial x_1} - x_1\frac{\partial}{\partial x_2}\right)V_3(x_1x_2x_3)$$

or in matrix notation,

$$\begin{bmatrix} V_1'(x_1'x_2'x_3') \\ V_2'(x_1'x_2'x_3') \\ V_3'(x_1'x_2'x_3') \end{bmatrix} = \left\{1 + \epsilon\left(x_2\frac{\partial}{\partial x_1} - x_1\frac{\partial}{\partial x_2}\right) + \epsilon\begin{bmatrix} 0 & -1 & 0 \\ 1 & 0 & 0 \\ 0 & 0 & 0 \end{bmatrix}\right\}\begin{bmatrix} V_1(x_1x_2x_3) \\ V_2(x_1x_2x_3) \\ V_3(x_1x_2x_3) \end{bmatrix}$$

Comparison with (5.40) shows that the curly bracket is $R = 1 - i\epsilon J_3$, and that J_3 is

$$J_3 = -i\left(x_1\frac{\partial}{\partial x_2} - x_2\frac{\partial}{\partial x_1}\right) + \begin{bmatrix} 0 & -i & 0 \\ i & 0 & 0 \\ 0 & 0 & 0 \end{bmatrix} \qquad (5.41)$$

The first part is the familiar x_3 component of the orbital angular momentum,

$$L_3 = -i\left(x_1\frac{\partial}{\partial x_2} - x_2\frac{\partial}{\partial x_1}\right)$$

the second part is the third component of the intrinsic spin of the vector field S_3. By considering infinitesimal rotations about x_1 and x_2, the other two components of the intrinsic spin $\mathbf{S}$ can be determined by cyclic permutation. The intrinsic spin operators for a vector field are, in the Cartesian representation,

$$S_1 = \begin{bmatrix} 0 & 0 & 0 \\ 0 & 0 & -i \\ 0 & i & 0 \end{bmatrix}, \quad S_2 = \begin{bmatrix} 0 & 0 & i \\ 0 & 0 & 0 \\ -i & 0 & 0 \end{bmatrix}, \quad S_3 = \begin{bmatrix} 0 & -i & 0 \\ i & 0 & 0 \\ 0 & 0 & 0 \end{bmatrix} \qquad (5.42)$$

The total angular momentum of a vector field is the vector sum of the orbital angular momentum $\mathbf{L} = -i\,\mathbf{r}\times\nabla$ and the intrinsic spin $\mathbf{S}$, the three Cartesian components of which are given in (5.42):

$$\mathbf{J} = \mathbf{L} + \mathbf{S} \qquad (5.43)$$

One can verify by direct multiplication of the matrices in (5.42) that the usual commutation rules for angular momentum operators are obeyed.

$$S_1 S_2 - S_2 S_1 = i S_3 \qquad \text{(c.p.)}$$

The eigenfunctions of the operators $\mathbf{L}^2$ and $\mathbf{L}_3$ are, of course, the spherical harmonics. Since the intrinsic spin operators (5.42) effect a reordering of the Cartesian components of the vector $\mathbf{V}$, it would appear that the eigenfunctions of $\mathbf{S}^2$ and S_3 are connected with the unit vectors $\mathbf{e}_1$, $\mathbf{e}_2$, and $\mathbf{e}_3$ along the Cartesian axes.

From (5.42) we see that the eigenvalue equation $S_3 \mathbf{V} = \mu \mathbf{V}$ becomes

$$S_3 \begin{bmatrix} V_1 \\ V_2 \\ V_3 \end{bmatrix} = \begin{bmatrix} -i V_2 \\ i V_1 \\ 0 \end{bmatrix} = \mu \begin{bmatrix} V_1 \\ V_2 \\ V_3 \end{bmatrix}$$

and $\mu = \pm 1, 0$. The solutions are

$$\mu = 1, \qquad \mathbf{V} = \pm V_1 (\mathbf{e}_1 + i \mathbf{e}_2)$$

$$\mu = 0, \qquad \mathbf{V} = V_3 \mathbf{e}_3$$

$$\mu = -1, \qquad \mathbf{V} = \pm V_1 (\mathbf{e}_1 - i \mathbf{e}_2)$$

with incoherent signs. These eigenfunctions are normalized by choosing $V_1 = 1/\sqrt{2}$. Also the phases are arbitrary. They are chosen so that the intrinsic spin eigenfunctions are just the spherical basis vectors

$$\boldsymbol{\xi}_1 = - \frac{1}{\sqrt{2}} (\mathbf{e}_1 + i \mathbf{e}_2)$$

$$\boldsymbol{\xi}_0 = \mathbf{e}_3 \qquad\qquad (5.44)$$

$$\boldsymbol{\xi}_1 = \frac{1}{\sqrt{2}} (\mathbf{e}_1 - i \mathbf{e}_2)$$

This is in complete analogy with the fact that the eigenfunctions for unit orbital angular momentum are Y_{1m} which, for $m = 1, 0, -1$ are proportional to

$$- \frac{1}{\sqrt{2}} (x_1 + i x_2)$$

$$x_3$$

$$\frac{1}{\sqrt{2}} (x_1 - i x_2)$$

respectively.

We will find it more convenient to transform the S-matrices, which appear in (5.42) in the Cartesian representation, to the representation of section 8 where S_3 is diagonal. This is done in much the same way as in section 15. However, it must be recognized that, whereas (4.41) indicates the manner of transformation for a vector, the matrix U, defined by (4.44), must be replaced by U^* when the spin matrices are transformed. This is seen by the argument that follows.

Let M be any matrix of three rows and columns. Then its matrix elements in the spherical representation are $M_{\mu\nu}^s$; μ, $\nu = \pm 1$, 0. Thus,

$$M \, \boldsymbol{\xi}_\nu = \sum_\mu M_{\mu\nu}^s \, \boldsymbol{\xi}_\mu \tag{5.45}$$

Since the spherical basis vectors are orthonormal according to

$$(-)^{\mu'} \, \boldsymbol{\xi}_{-\mu'} \cdot \boldsymbol{\xi}_\mu = \boldsymbol{\xi}_{\mu'}^* \cdot \boldsymbol{\xi}_\mu = \delta_{\mu'\mu} \tag{5.46}$$

it follows that

$$M_{\mu\nu}^s = \boldsymbol{\xi}_\mu^* \cdot M \, \boldsymbol{\xi}_\nu \tag{5.47}$$

In a similar way the Cartesian representation is defined by

$$M_{ij} = \mathbf{e}_i \cdot M \mathbf{e}_j \tag{5.48}$$

But, by definition, U transforms the Cartesian basis into the spherical basis. That is,

$$\boldsymbol{\xi}_\mu = \sum_i U_{\mu i} \mathbf{e}_i \tag{5.49}$$

Combining (5.47), (5.48), and (5.49), we find

$$M_{\mu\nu}^s = \sum_{ij} U_{\mu i}^* U_{\nu j} M_{ij} = (U^* M \tilde{U})_{\mu\nu} \tag{5.50}$$

where the tilde means transpose. Alternatively, since U is unitary, this becomes

$$M^s = \tilde{U}^{-1} M \tilde{U} = U^* M U^{*-1} \tag{5.51}$$

which corresponds to replacing U by U^* in (4.41), as was stated above.

The diagonal S_3 representation of the intrinsic spin operators $\mathbf{S}$ (S_1, S_2, S_3) is $\mathbf{S}'$ where

$$\mathbf{S}' = U^* \mathbf{S} U^{*-1}$$

In detail this gives

$$S_1' = \frac{1}{\sqrt{2}} \begin{bmatrix} 0 & 1 & 0 \\ 1 & 0 & 1 \\ 0 & 1 & 0 \end{bmatrix}, \qquad S_2' = \frac{1}{\sqrt{2}} \begin{bmatrix} 0 & -i & 0 \\ i & 0 & -i \\ 0 & i & 0 \end{bmatrix}$$

$$S_3' = \begin{bmatrix} 1 & 0 & 0 \\ 0 & 0 & 0 \\ 0 & 0 & -1 \end{bmatrix} \tag{5.52}$$

In this representation the spherical basis components are

$$S_1 = -\frac{1}{\sqrt{2}}(S_1' + iS_2') = \begin{bmatrix} 0 & -1 & 0 \\ 0 & 0 & -1 \\ 0 & 0 & 0 \end{bmatrix}$$

$$S_0 = S_3' = \begin{bmatrix} 1 & 0 & 0 \\ 0 & 0 & 0 \\ 0 & 0 & -1 \end{bmatrix}$$

$$S_{-1} = \frac{1}{\sqrt{2}}(S_1' - iS_2') = \begin{bmatrix} 0 & 0 & 0 \\ 1 & 0 & 0 \\ 0 & 1 & 0 \end{bmatrix} \qquad (5.53)$$

That is, $(S_{\pm 1})_{\mu'\mu} = \mp\delta_{\mu,\mu'\pm 1}$, $(S_0)_{\mu\mu'} = \mu\,\delta_{\mu\mu'}$
In any representation

$$\mathbf{S}^2 = S_1^2 + S_2^2 + S_3^2 = 2\begin{bmatrix} 1 & 0 & 0 \\ 0 & 1 & 0 \\ 0 & 0 & 1 \end{bmatrix} \qquad (5.54)$$

is diagonal, and $S(S+1) = 2$, giving $S = 1$. Therefore the intrinsic spin of a vector field is unity. This conclusion applies, of course, to the electromagnetic field so that the intrinsic spin of the photon is also unity.

Since both $\mathbf{S}^2$ and S_0 are diagonal in the spherical basis representation (5.53), we verify that the eigenfunctions are $\boldsymbol{\xi}_\mu$ with $\mu = \pm 1, 0$. Indeed, from the preceding it may be seen that

$$S_\mu\,\boldsymbol{\xi}_\nu = (-)^\mu\sqrt{2}\,C(111;\mu+\nu,-\mu)\,\boldsymbol{\xi}_{\mu+\nu} \qquad (5.55)$$

which is equivalent to (5.13) with $j = 1$.

It is also clear that the components of $\mathbf{S}$ constitute an irreducible tensor of rank one, as is the case for angular momentum operators. This is clear from (5.13) and the Wigner–Eckart theorem. On the other hand, $\mathbf{S}^2$, or the square of any angular momentum operator, is a zero-rank tensor. This is so since, for any two vectors $\mathbf{A}$ and $\mathbf{B}$,

$$\mathbf{A} = \sum_\mu (-)^\mu A_\mu\,\boldsymbol{\xi}_{-\mu}; \qquad A_\mu = \mathbf{A}\cdot\boldsymbol{\xi}_\mu$$

and

$$\mathbf{A}\cdot\mathbf{B} = \sum_\mu (-)^\mu A_\mu\,B_{-\mu} \qquad (5.56)$$

$\mathbf{S}^2$ or $\mathbf{J}^2$ is obtained by setting $\mathbf{A} = \mathbf{B} = \mathbf{S}$ or $\mathbf{J}$. From (5.3) it is clear that, as expected, (5.56) represents an invariant, or a zero-rank tensor.

The eigenfunctions of the total angular momentum are those that diagonalize $\mathbf{J}^2$ and J_z. These are now very easy to construct. From what has been said they are

$$T_{JM} = \sum_m C(L1J; M-m, m) \, Y_{L,M-m}(\mathbf{r}) \, \boldsymbol{\xi}_m \qquad (5.57)$$

Thus

$$\mathbf{J}^2 \, \mathbf{T}_{JM} = J(J+1) \, \mathbf{T}_{JM} \qquad (5.58a)$$

$$J_z \, \mathbf{T}_{JM} = M \, \mathbf{T}_{JM} \qquad (5.58b)$$

$$\mathbf{L}^2 \, \mathbf{T}_{JM} = L(L+1) \, \mathbf{T}_{JM} \qquad (5.58c)$$

$$\mathbf{S}^2 \, \mathbf{T}_{JM} = 2\mathbf{T}_{JM} \qquad (5.58d)$$

and the $\mathbf{T}_{JM}$ constitute an irreducible tensor [12] of rank J. They depend parametrically on L, and for given J there are three possible L values; $L = J, J\pm1$. Neither J nor L may be negative.

The projections $\mathbf{T}_{JM} \cdot \boldsymbol{\xi}_m^*$ are also irreducible tensors but of rank L. On the other hand the scalar product of the tensor $\mathbf{T}_{JM}$ and the vector $\mathbf{a}$ is

$$\mathbf{T}_{JM} \cdot \mathbf{a} = \left(\frac{4\pi}{3}\right)^{\frac{1}{2}} \sum_m C(L1J; M-m, m) \, Y_{L,M-m}(\mathbf{r}) \, \mathcal{Y}_{1m}(\mathbf{a}) \quad (5.59)$$

where $\mathcal{Y}_{1m}(\mathbf{a})$ is the mth component of the first-degree solid harmonic of argument $\mathbf{a}$, and this scalar product is an irreducible tensor of rank J provided that $\mathbf{a}$ is an irreducible tensor of rank one. The discussion at the end of section 17 should be consulted in this connection.

These tensors are orthonormal on the surface of the unit sphere.[4] Thus, introducing the dependence on L explicitly

$$\int \mathbf{T}_{JLM}^* \cdot \mathbf{T}_{J'L'M'} \, d\Omega = \delta_{JJ'} \, \delta_{LL'} \, \delta_{MM'} \qquad (5.60)$$

This follows immediately from the definition (5.57) and the unitary property of the C-coefficients. On the surface of the unit sphere these tensors must form a complete set. They are of particular interest insofar as they are the spin-angular eigenfunctions of the Maxwell field; see sections 26 and 27.

VI. RACAH COEFFICIENTS

22. COUPLING OF THREE ANGULAR MOMENTA

In Chapter III we discussed the coupling of two angular momenta j_1 and j_2 to give a resultant $j = j_1 + j_2$, and also the unitary transformation from the representation in which j_1^2, j_{1z} and j_2^2, j_{2z} are diagonal to the representation in which j^2, j_z and j_1^2, j_2^2 are diagonal. The elements of this unitary transformation are the Clebsch–Gordan coefficients $C(j_1 j_2 j; m_1 m_2 m)$, and much of Chapter III was devoted to investigating their properties. We now consider the addition of three angular momenta j_1, j_2, j_3 to form the total angular momentum $j = j_1 + j_2 + j_3$, and we shall study the coefficients of the unitary transformations connecting different representations of these angular momenta.[1] The necessity for this generalization arises from the fact that this simple type of coupling occurs in almost every problem of physical interest. For example, attention is drawn to the emission or absorption of radiation. The unitary transformations referred to represent a recoupling since they connect different ways of forming the resultant from the three addends. As a matter of nomenclature, the elements of this unitary matrix are, within a multiplicative factor [see (6.3) below], the Racah coefficients which are of prime importance in the theory of angular momentum. A fairly detailed account will also be given of the properties of these Racah coefficients.[1]

As implied above, for three angular momenta there are several sets of commuting operators which may be diagonalized simultaneously. Each set contains six operators, as may be seen from the uncoupled representation, in which $j_1^2, j_2^2, j_3^2, j_{1z}, j_{2z}, j_{3z}$ are diagonal. If a representation is desired in which the square and z-component of the total angular momentum are diagonal, this can be obtained by compounding any two of the angular momenta into an intermediate angular momentum, and then compounding this intermediate angular momentum with the third member of the original set. Thus, the six operators which may be diagonalized are

$$j_1^2, j_2^2, j_3^2; j_{int}^2, j^2, j_z$$

[1] G. Racah, *Phys. Rev.* **62**, 438 (1942); **63**, 367 (1943).

There are three such representations since j_{int} can be either $j_1 + j_2$, $j_2 + j_3$, or $j_1 + j_3$. Let us consider the connection between the first two representations characterized by the intermediate angular momenta

$$j' = j_1 + j_2, \qquad j'' = j_2 + j_3 \qquad (6.1)$$

the corresponding eigenstates [2] being $\psi_{jm}(j')$ and $\psi_{jm}(j'')$. These two representations are related by a unitary transformation, the elements of which we write as $R_{j''j'}$:

$$\psi_{jm}(j') = \sum_{j''} R_{j''j'} \, \psi_{jm}(j'') \qquad (6.2)$$

The Racah coefficients W are then defined by the following relation:

$$R_{j''j'} = [(2j'' + 1)(2j' + 1)]^{\frac{1}{2}} \, W(j_1 j_2 j j_3; j' j'') \qquad (6.3)$$

The two types of eigenfunctions in (6.1) can easily be written down from the rules for coupling two angular momenta given in Chapter III. In this way

$$\psi_{j'm'} = \sum_{m_1} C(j_1 j_2 j'; m_1, m'-m_1) \, \psi_{j_1 m_1} \psi_{j_2 \, m'-m}.$$

is an eigenfunction of $\mathbf{j}'^2$ and j_z' (as well as $\mathbf{j}_1^2$ and $\mathbf{j}_2^2$) and

$$\psi_{jm}(j') = \sum_{m'} C(j'j_3 j; m', m-m') \, \psi_{j_3 \, m-m'}$$
$$\times \sum_{m_1} C(j_1 j_2 j'; m_1, m'-m_1) \, \psi_{j_1 m_1} \psi_{j_2 \, m'-m_1}$$

is an eigenfunction of $\mathbf{j}^2$ and j_z (as well as $\mathbf{j}'^2$ and $\mathbf{j}_3^2$). Similarly,

$$\psi_{j''m''} = \sum_{m_2} C(j_2 j_3 j''; m_2, m''-m_2) \, \psi_{j_2 m_2} \psi_{j_3 \, m''-m_2}$$

is an eigenfunction of $\mathbf{j}''^2$ and j_z (as well as $\mathbf{j}_2^2$ and $\mathbf{j}_3^2$) and

$$\psi_{jm}(j'') = \sum_{m} C(j_1 j''j; m-m'', m'') \, \psi_{j_1 \, m-m''}$$
$$\times \sum_{m_2} C(j_2 j_3 j''; m_2, m''-m_2) \, \psi_{j_2 m_2} \psi_{j_3 \, m''-m_2}$$

is an eigenfunction of $\mathbf{j}^2$ and j_z (as well as $\mathbf{j}_1^2$ and $\mathbf{j}''^2$). Substitution of these wave functions into (6.3) gives

$$\sum_{m_1} \sum_{m'} C(j_1 j_2 j'; m_1, m'-m_1) \, C(j'j_3 j; m', m-m') \, \psi_{j_1 m_1} \psi_{j_2 \, m'-m_1} \psi_{j_3 \, m-m'}$$
$$= \sum_{m_2} \sum_{m''} \sum_{j''} R_{j''j'} \, C(j_1 j''j; m-m'', m'') \, C(j_2 j_3 j''; m_2, m''-m_2)$$
$$\times \psi_{j_1 \, m-m''} \psi_{j_2 m_2} \psi_{j_3 \, m''-m_2}$$

---

[2] Once more the same symbol is used for eigenfunctions in different spaces. The intermediate angular momentum which characterizes the representation is sufficient for purposes of notation. The redundant quantum numbers j_1, j_2, j_3, have also been suppressed.

If we take the scalar product of both sides of this equation with $\psi_{j_1\mu_1}\psi_{j_2\mu_2}\psi_{j_3\mu_3}$, it becomes

$$\sum_{m_1}\sum_{m'} C(j_1j_2j'; m_1, m'-m_1)\, C(j'j_3j; m', m-m')\, \delta_{m_1\mu_1}\, \delta_{m'-m_1,\mu_2}\, \delta_{m-m',\mu_3}$$

$$= \sum_{m_2}\sum_{m''}\sum_{j''} R_{j''j'}\, C(j_1j''j; m-m'', m'')\, C(j_2j_3j''; m_2, m''-m_2)$$
$$\times\, \delta_{m-m'',\mu_1}\, \delta_{m_2,\mu_2}\, \delta_{m''-m_2,\mu_3}$$

or

$$C(j_1j_2j'; \mu_1\mu_2)\, C(j'j_3j; \mu_1+\mu_2, \mu_3)\, \delta_{\mu_1+\mu_2+\mu_3,m}$$

$$= \sum_{j''} R_{j''j'}\, C(j_1j''j; \mu_1, m-\mu_1)\, C(j_2j_3j''; \mu_2, m-\mu_1-\mu_2)\, \delta_{\mu_1+\mu_2+\mu_3,m}$$

The δ symbols can be eliminated by replacing m with $\mu_1+\mu_2+\mu_3$:

$$C(j_1j_2j'; \mu_1\mu_2)\, C(j'j_3j; \mu_1+\mu_2, \mu_3)$$
$$= \sum_{j''} R_{j''j'}\, C(j_2j_3j''; \mu_2\mu_3)\, C(j_1j''j; \mu_1, \mu_2+\mu_3) \quad (6.4a)$$

Equivalent relations can be obtained by use of the orthonormality of the C-coefficients. Multiplication by $C(j_2j_3J; \mu_2\mu_3)$ and summation over μ_2, keeping $\mu_2+\mu_3$ fixed, yields this result,

$$\sum_{\mu_2} C(j_1j_2j'; \mu_1\mu_2)\, C(j'j_3j; \mu_1+\mu_2, \mu_3)\, C(j_2j_3j''; \mu_2\mu_3)$$
$$= R_{j''j'}\, C(j_1j''j; \mu_1, \mu_2+\mu_3) \quad (6.5a)$$

after the extra prime is dropped. Similarly, multiplication by $C(j_1j''J; \mu_1, \mu_2+\mu_3)$ and summation over μ_1, keeping the $\mu_1+\mu_2+\mu_3$ fixed, gives (after the J is changed to j)

$$\sum_{\mu_1}\sum_{\mu_2} C(j_1j_2j'; \mu_1\mu_2)\, C(j'j_3j; \mu_1+\mu_2, \mu_3)$$
$$\times\, C(j_1j''j; \mu_1, \mu_2+\mu_3)\, C(j_2j_3j''; \mu_2\mu_3) = R_{j''j'} \quad (6.6a)$$

The elements of the unitary matrix R, and hence W, do not depend on the eigenvalues of j_z or on any other magnetic quantum number. Thus, in (6.6a) both sides are independent [3] of μ_3. From (6.2) it is seen

[3] Had we considered the third coupling possibility

$$j_1 + j_3 = j''' \quad \text{and} \quad j''' + j_2 = j$$

nothing essentially new would have been introduced. Designating the wave functions for this coupling scheme by $\psi_{jm}(j''')$ and writing

$$\psi_{jm}(j''') = \sum_{j''} S_{j''j'''}(j_1j_2j_3j)\, \psi_{jm}(j'')$$

we would find

$$S_{j''j'''}(j_1j_2j_3j) = (-)^{j_2+j_3-j''}\, R_{j''j'''}(j_1j_3j_2j)$$

Here we have made the notation more explicit by recognizing that the R and S matrices depend on six angular momentum values and $R_{j''j'}(j_1j_2j_3j)$ is given by (6.6a).

that $R_{j''j'}$ is independent of m since both representations belong to the same total projection quantum number. The foregoing statement then follows when we remember that $m = \mu_1 + \mu_2 + \mu_3$.

23. PROPERTIES OF THE RACAH COEFFICIENTS

We have defined the Racah coefficient $W(j_1j_2jj_3; j'j'')$ as $[(2j' + 1)(2j'' + 1)]^{-\frac{1}{2}}$ times the matrix element of the unitary transformation $R_{j''j'}$. This definition is not really complete, however, until an explicit formula is given for these coefficients. In general this implies some phase conventions for the eigenfunctions $\psi_{j_1m_1}$, etc. Actually, this has already been done in the process of making the C-coefficients explicit so that the W-coefficients are completely defined. Equations (6.4a), (6.5a), and (6.6a) represent such defining expressions since they relate the Racah coefficients to the known and completely defined Clebsch–Gordan coefficients. It is clear that W's are all real. In the following we shall follow Racah's notation of using the lower-case Latin letters $abcd$; ef for the angular momenta $j_1j_2jj_3$; $j'j''$ and lower-case Greek letters for the magnetic quantum numbers associated with the first four angular momenta: $\alpha\beta\gamma\delta$ ($\gamma = \alpha + \beta + \delta$). The above relations are then

$$\sum_f [(2e + 1)(2f + 1)]^{\frac{1}{2}} W(abcd; ef) \, C(bdf; \beta\delta) \, C(afc; \alpha, \beta+\delta)$$

$$= C(abe; \alpha\beta) \, C(edc; \alpha+\beta, \delta) \quad (6.4b)$$

$$[(2e + 1)(2f + 1)]^{\frac{1}{2}} W(abcd; ef) \, C(afc; \alpha, \beta+\delta)$$

$$= \sum_\beta C(abd; \alpha\beta) \, C(edc; \alpha+\beta, \delta) \, C(bdf; \beta\delta) \quad (6.5b)$$

$$[(2e + 1)(2f + 1)]^{\frac{1}{2}} W(abcd; ef)$$

$$= \sum_\alpha \sum_\beta C(abe; \alpha\beta) \, C(edc; \alpha+\beta, \delta) \, C(bdf; \beta\delta) \, C(afc; \alpha, \beta+\delta) \quad (6.6b)$$

Note that on the left in (6.5b) the sum $\beta + \delta = \gamma - \alpha$ which is a free parameter.

Following Racah,[1] we can use (6.6b) as the definition of the W-coefficient. Using algebraic expressions for the Clebsch–Gordan coefficients, it is possible to carry out the double sum in (6.6b) and to obtain the following closed expression for the Racah coefficients:

$$W(abcd; ef) = \Delta_R(abe) \, \Delta_R(cde) \, \Delta_R(acf) \, \Delta_R(bdf)$$

$$\times \sum_\varkappa \frac{(-)^{\varkappa+a+b+c+d}(\varkappa + 1)!}{(\varkappa - a - b - e)!(\varkappa - c - d - e)!(\varkappa - a - c - f)!(\varkappa - b - d - f)!}$$

$$\times \frac{1}{(a + b + c + d - \varkappa)!(a + d + e + f - \varkappa)!(b + c + e + f - \varkappa)!}$$

$$(6.7)$$

The Δ_R is the "triangle" coefficient, symmetric in its arguments,

$$\Delta_R(abc) = \left[\frac{(a+b-c)!(a-b+c)!(-a+b+c)!}{(a+b+c+1)!} \right]^{\frac{1}{2}} \quad (6.8)$$

This function vanishes unless the usual triangular condition in a, b, and c is satisfied. The triangle coefficients in (6.7) contain the rules for the non-vanishing of the Racah coefficients. These can also be deduced from the definition (6.6a) in terms of sums of products of four Clebsch–Gordan coefficients. Since only the projection quantum numbers change from term to term, the conditions for the non-vanishing of the four Clebsch–Gordan coefficients are expressed by

$$W(j_1 j_2 j j_3; j'j'') = 0 \quad \text{unless} \quad \begin{cases} \Delta(j_1 j_2 j'), \ \Delta(j'j_3 j) \\ \Delta(j_2 j_3 j''), \ \Delta(j_1 j''j) \end{cases} \quad (6.9a)$$

Of course, these are just the rules for coupling the various angular momenta which define the two coupling schemes, $\psi_{jm}(j')$ and $\psi_{jm}(j'')$, and so their appearance is no surprise but is, in fact, a necessity. In the alphabetical notation, the condition (6.9a) is

$$W(abcd; ef) = 0 \quad \text{unless} \quad \Delta(abe), \ \Delta(edc), \ \Delta(bdf), \ \Delta(afc) \quad (6.9b)$$

Each of the triangles involves two angular momenta to the left and one to the right of the semicolon in $W(abcd; ef)$. These four relations are conveniently visualized by forming a quadrilateral from the four triangles whose sides are the angular momenta to the left of the semicolon and whose diagonals are the two angular momenta to the right of the semicolon.

The roles of the six angular momenta in $W(abcd; ef)$ may be interchanged provided the four triangular relations are preserved. As expected, interchange of the parameters, with (6.9b) still valid, results in a phase change at most.

$$W(abcd; ef) = W(badc; ef) = W(cdab; ef) = W(acbd; fe) \quad (6.10a)$$

$$W(abcd; ef) = (-)^{e+f-a-d} W(ebcf; ad) = (-)^{e+f-b-c} W(aefd; bc) \quad (6.10b)$$

In all there are twenty-four possible arrangements of the six arguments wherein the triangular relations are preserved. The relations between the Racah coefficients are given more completely in Appendix I. If we refer to the quadrilateral in Fig. 7a, we see that the Racah coefficient is invariant under interchanges of sides with sides (which occurs in pairs) and a diagonal with a diagonal. On the other hand, the interchange of a side with a diagonal can only be done in pairs (both diagonals must interchange with sides), and this always requires a phase change. To determine whether a sign change is introduced, we add the numbers that cross the semicolon to the left under such a permutation, $e + f$, and then subtract the numbers that cross the semicolon to the right,

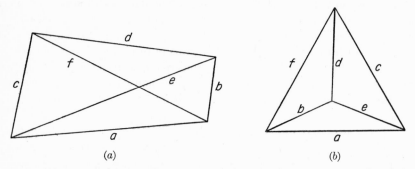

(a)   (b)

Fig. 7. The quadrilateral of Fig. 7a illustrates the relation of the six vectors forming the arguments of a Racah coefficient. An equivalent (tetrahedral) arrangement is shown in Fig. 7b.

$a + d$ or $b + c$ in (6.10b). By the triangular rules $e + f - a - d$ and $e + f - b - c$ are always integers. There is a sign change if these sums are odd. By means of these symmetry relations, it is always possible to bring any one of the arguments into a given position. An equivalent geometrical construction is given in Fig. 7b. Here the same four triangles are represented, but in this tetrahedral structure all six parameters play essentially equivalent roles. This suggests that the phases introduced in the interchange of these parameters could be removed by a suitable redefinition. Indeed,

$$T(abcd; ef) = (-)^{a+b+c+d} W(abcd; ef)$$

is symmetrical with respect to any parameter interchange which preserves the four triangular relations.

When the area of one of the triangles vanishes, that is, when one member of a triad in (6.9b) equals the sum of the other two, the sum in (6.7) reduces to a single term. For example, suppose $e = a + b$. Then one of the factors in the $\varkappa$ term of the sum is

$$\frac{1}{(\varkappa - a - b - c - d)!(a + b + c + d - \varkappa)!}$$

Since all the arguments of W are non-negative, the restriction of $\varkappa$ to integral values implies that $\varkappa \geq a + b + c + d$ and $\varkappa \leq a + b + c + d$.

This can be so for only one value of $\varkappa = a + b + c + d$. In this way we find that

$$
W(abcd; a+b, f) = \left[\frac{(2a)!(2b)!}{(2a + 2b + 1)!} \right.
$$

$$
\times \frac{(a + b + c - d)!(a + b + c + d + 1)!(a + b + d - c)!}{(a + c + f + 1)!(c + d - a - b)!(b + f + d + 1)!}
$$

$$
\times \left. \frac{(c + f - a)!(d + f - b)!}{(a + c - f)!(a + f - c)!(b + d - f)!(b + f - d)!} \right]^{\frac{1}{2}} \quad (6.11)
$$

By use of the symmetry relations (6.10) this equation can be used for the general case where any one of the four triangles in (6.9) has zero area. Another example of a triangle with vanishing area occurs when one of the arguments vanishes, such as in $W(abcd; 0f)$. In this case both the triangles $\Delta(abe)$ and $\Delta(edc)$ have vanishing area, and the Racah coefficient vanishes unless $a = b$ and $c = d$. Use of the symmetry relation (6.10b) changes $W(abcd; 0f)$ into the Racah coefficient

$$
(-)^{f-a-d} \, W(0bcf; ad) = (-)^{f-a-d} \, \delta_{ab} \, \delta_{cd} \, W(0bcf; bc)
$$

The expression in (6.11) can now be used if the arguments there are transformed as follows: $a \to 0, b \to b, c \to c, d \to f, f \to c$. In this way we find that

$$
W(abcd; 0f) = \frac{(-)^{f-b-d} \, \delta_{ab} \, \delta_{cd}}{[(2b + 1)(2d + 1)]^{\frac{1}{2}}} \quad (6.12)
$$

Since the Racah coefficients $W(abcd; ef)$ were defined in terms of the elements R_{fe} of the unitary transformation from the orthonormal eigenfunctions $\psi_{jm}(f)$ to the orthonormal eigenfunctions $\psi_{jm}(e)$, it follows that we can immediately deduce their orthonormality properties. The coefficients R_{fe} must therefore satisfy the orthonormality conditions

$$
\sum_e R_{fe}R_{ge} = \delta_{fg} \quad \text{and} \quad \sum_e R_{ef}R_{eg} = \delta_{fg}
$$

Substitution of (6.3) into these equations gives the orthonormality property of the Racah coefficients

$$
\sum_e (2e + 1)(2f + 1) \, W(abcd; ef) \, W(abcd; eg) = \delta_{fg} \quad (6.13)
$$

Although there are two orthonormality relations for the R's, only one has been written for the W's since the second can be obtained from it with the symmetry rules (6.10).

Another important sum over W-coefficients is Racah's sum rule [1]

$$\sum_e (2e + 1)(-)^{a+b-e} W(abcd; ef) W(bacd; eg) = W(agfb; cd) \quad (6.14)$$

By considering the coupling of four angular momenta, Biedenharn [4] was able to derive another sum rule for the Racah coefficients

$$\sum_g (2g + 1) W(a'gdc; ac') W(bgec'; b'c) W(a'gfb; ab')$$
$$= W(adbe; cf) W(a'db'e; c'f) \quad (6.15)$$

Racah has shown that, except for a phase, the symmetry relations (6.10), the orthonormality property (6.14), and the two sum rules (6.14) and (6.15) completely determine the Racah coefficients.

The above completes our definition and discussion of the formal properties of the Racah coefficient $W(abcd; ef)$. This material is covered in greater detail in the review article of Biedenharn, Blatt, and Rose [5] and in the tables of Biedenharn.[6] The first-mentioned reference includes algebraic expressions for the cases $e = \frac{1}{2}$, 1, $\frac{3}{2}$, 2, which were obtained from (6.7). For convenience the tables for $e = \frac{1}{2}$ and 1 are reproduced in Appendix I. Jahn [7] has given the same formulae, but for the related unitary matrix element R_{fe} of (6.2). Lloyd [8] and Biedenharn [6] have obtained some useful specializations of equations (6.4b), (6.5b), and (6.6b), and other identities relating the Racah and Clebsch–Gordan coefficients. Because the algebraic formulae for Racah coefficients become extremely complicated for large arguments, Biedenharn [6] prepared a numerical table of these coefficients for the following ranges of arguments: $e = \frac{1}{2}$, 1, $\frac{3}{2}$, 2, $\frac{5}{2}$, 3; $f = 0, 1, 2, 3, 4, 5, 6, 7, 8$; $a,c = 0, 1, 2, 3, 4$; and $b,d \le 4$. Values of W are square roots, with appropriate sign, of the rational fractions appearing therein. An extended table in decimal form has since been computed by Simon, Vander Sluis, and Biedenharn.[9] The ranges of arguments of this tabulation are the same as in Biedenharn's,[6] except

[4] L. C. Biedenharn, *J. Math. Phys.* **31**, 287 (1953).

[5] L. C. Biedenharn, J. M. Blatt, and M. E. Rose, *Revs. Mod. Phys.* **24**, 249 (1952).

[6] L. C. Biedenharn, *Oak Ridge Natl. Lab. Rept. 1098*. This report also contains an extensive table of Z-coefficients. These coefficients, defined by

$$Z(abcd; ef) = i^{f-a+c}[(2a + 1)(2b + 1)(2c + 1)(2d + 1)]^{\frac{1}{2}} C(acf; 00) W(abcd; ef)$$

are important in the description of angular distributions of particles with intrinsic spin 0 and $\frac{1}{2}$; see Chapters X and XI.

[7] H. A. Jahn, *Proc. Roy. Soc. London* **A 205**, 235 (1951).

[8] S. P. Lloyd, Ph.D. thesis, University of Illinois (1951). See also M. E. Rose, L. C. Biedenharn, and G. B. Arfken, *Phys. Rev.* **85**, 5 (1952).

[9] A. Simon, J. H. Vander Sluis, L. C. Biedenharn, *Oak Ridge Natl. Lab. Rept. 1679* (1954).

that a and c are no longer restricted to integral values. Additional tabulation of the Racah coefficients has been carried out by Obi et al.[10] These tables are quite extensive and appear in three parts. For example, part I which presents values of the coefficients for integral parameters applies to all cases for which e and $f \leq 7$. Parts II and III contain coefficients with three and four half-integers. No other possibilities can occur. These all appear as square roots of rational fractions. Associated functions, which occur in angular distributions of particles of spin ≤ 1, have been tabulated by Sharp et al.[11]

24. BASIC APPLICATIONS

The first application of Racah coefficients that we consider is the matrix elements of the contraction of two irreducible tensors of rank L. This is written as

$$\mathbf{T}_L(1) \cdot \mathbf{T}_L(2) = \sum_M (-)^M T_{LM}(1) \, T_{L,-M}(2) \qquad (6.16)$$

We have already shown in section 17 that such a contraction is a tensor of zero rank, which means that it is invariant under rotations. Since interaction energies are invariants, they can be written in general as the sum of such invariants. Usually the system can be divided into two distinct parts, so that one tensor depends exclusively on one set of variables and the other tensor on a set of variables in a different space. The two sets of variables are indicated by the arguments 1 and 2 in (6.16). One can often characterize the two components of the system by their angular momenta $\mathbf{j}_1$ and $\mathbf{j}_2$. The problem then arises of evaluating the matrix elements of (6.16) in the coupled representation wherein the angular momenta $\mathbf{j}_1$ and $\mathbf{j}_2$ are compounded into the resultant $\mathbf{j}$. As an example, we cite the matrix elements of the spin-orbit coupling $\mathbf{L \cdot s}$ in the representation in which $\mathbf{j}^2$ and j_z are diagonal and $\mathbf{j} = \mathbf{L} + \mathbf{s}$; we shall therefore consider

$$(j'm'j_1'j_2' | \mathbf{T}_L(1) \cdot \mathbf{T}_L(2) | jmj_1j_2) \qquad (6.17)$$

This is the matrix elements of $\mathbf{T}_L(1) \cdot \mathbf{T}_L(2)$ between the state

$$\psi_{j'm'} = \sum_{m_1'} C(j_1'j_2'j'; m_1', m'-m_1') \, \psi_{j_1'm_1'} \psi_{j_2' \, m'-m_1'} \qquad (6.18a)$$

and the state

$$\psi_{jm} = \sum_{m_1} C(j_1j_2j; m_1, m-m_1) \, \psi_{j_1m_1} \psi_{j_2 \, m-m_1} \qquad (6.18b)$$

[10] S. Obi, T. Ishidzu, H. Horie, S. Yanagowa, Y. Tanabe, and M. Sato, *Ann. Tokyo Astron. Observ.*, Second Ser., **III**, no. 3, 89 (1953); **IV**, no. 1, 1 (1954); **IV**, no. 2, 77 (1955).

[11] W. T. Sharp, J. M. Kennedy, B. J. Sears, and M. G. Hoyle, *Chalk River Rept. CRT-556.*

Substitution of equations (6.18) and (6.16) into (6.17) yields

$$\sum_M \sum_{m_1'} \sum_{m_1} (-)^M \, C(j_1'j_2'j'; m_1', m'-m_1') \, C(j_1j_2j; m_1, m-m_1)$$

$$\times \, (\psi_{j_1'm_1'} \, \psi_{j_2' \, m'-m_1'} | T_{LM}(1) \, T_{L,-M}(2) | \psi_{j_1m_1} \, \psi_{j_2 \, m-m_1})$$

The matrix element can be split into two factors

$$(j_1'm_1' | T_{LM}(1) | j_1m_1)(j_2' \, m'-m_1' | T_{L,-M}(2) | j_2 \, m-m_1)$$

and the Wigner–Eckart theorem (5.14) can be applied. Then we have

$$(j'm'j_1'j_2' | T_L(1) \cdot T_L(2) | jmj_1j_2) = (j_1' \parallel T_L(1) \parallel j_1)(j_2' \parallel T_L(2) \parallel j_2)$$

$$\times \sum_M \sum_{m_1'} \sum_{m_1} (-)^M \, C(j_1'j_2'j'; m_1', m'-m_1') \, C(j_1j_2j; m_1, m-m_1)$$

$$\times \, C(j_1Lj_1'; m_1Mm_1') \, C(j_2Lj_2'; m-m_1, -M, m'-m_1')$$

If we apply the Wigner–Eckart theorem directly to (6.17), and use the fact that $T_L(1) \cdot T_L(2)$ is a tensor of zero rank,

$$(j'm' | T_L(1) \cdot T_L(2) | jm) = C(j0j'; m0m')(j' \parallel T_L(1) \cdot T_L(2) \parallel j)$$

$$= \delta_{mm'} \, \delta_{jj'}(j \parallel T_L(1) \cdot T_L(2) \parallel j) \qquad (6.19)$$

we see that the matrix element under consideration vanishes unless $m' = m$ and $j' = j$. In addition, one of the sums can be eliminated by setting $M = m_1' - m_1$. Consequently,

$$(j'm'j_1'j_2' | T_L(1) \cdot T_L(2) | jmj_1j_2)$$

$$= \delta_{m'm} \, \delta_{j'j}(j_1' \parallel T_L(1) \parallel j_1)(j_2' \parallel T_L(2) \parallel j_2)$$

$$\times \sum_{m_1'} \sum_{m_1} (-)^{m_1'-m_1} \, C(j_1j_2j; m_1, m-m_1) \, C(j_2Lj_2'; m-m_1, m_1-m_1')$$

$$\times \, C(j_1Lj_1'; m_1, m_1'-m_1) \, C(j_1'j_2'j; m_1', m-m_1') \qquad (6.20)$$

Equations (6.4), (6.5), and (6.6) provide several alternatives for evaluating this double sum. However, the first of these relations is the one we shall use. To use (6.4b), there must be two C-coefficients of the following type: $C(abe; \alpha\beta) \, C(edc; \alpha+\beta, \delta)$. That is: (1) there must be a common argument e which occupies the third place in the first coefficient and the first place in the second coefficient; (2) the magnetic quantum numbers associated with e must be the same in the two coefficients. Finally, either β and δ or α and $\delta + \beta$ must be independent of the projection quantum number over which a summation is to be first performed. In that case one of the C-coefficients will not depend on the summation parameter. These conditions are satisfied by the last two Clebsch–Gordan coefficients in our sum, which now becomes

$$\sum_{m_1} \sum_{m_1'} (-)^{m_1'-m_1} C(j_1 j_2 j; m_1, m-m_1) \, C(j_2 L j_2''; m-m_1, m_1-m_1')$$
$$\times \sum_f [(2j_1'+1)(2f+1)]^{\frac{1}{2}} \, W(j_1 L j j_2''; j_1' f)$$
$$\times C(L j_2' f; m_1'-m_1, m-m_1') \, C(j_1 f j; m_1, m-m_1)$$

If we next sum over m_1', it is apparent that one of the new C-coefficients, as well as one of the original four, does not depend on m_1' at all, so that the sum over m_1' is just

$$\sum_{m_1'} (-)^{m_1'-m_1} C(j_2 L j_2''; m-m_1, m_1-m_1') \, C(L j_2' f; m_1'-m_1, m-m_1')$$

Using the symmetry relation (3.17c) in the first C-coefficient, we find for this latter sum

$$(-)^L \left[\frac{2j_2''+1}{2j_2+1} \right]^{\frac{1}{2}} \sum_{m_1'} C(L j_2'' j_2; m_1'-m_1, m-m_1') \, C(L j_2' f; m_1'-m_1, m-m_1')$$

and then, by the orthonormality of the C-coefficients, this is

$$(-)^L \left[\frac{2j_2''+1}{2j_2+1} \right]^{\frac{1}{2}} \delta_{j_2 f}$$

The remaining sum over m_1 just gives unity,

$$\sum_{m_1} C(j_1 j_2 j; m_1, m-m_1) \, C(j_1 j_2 j; m_1, m-m_1) = 1$$

so that the original sum over four Clebsch–Gordan coefficients is just $(-)^L [(2j_1'+1)(2j_2''+1)]^{\frac{1}{2}} W(j_1 L j j_2''; j_1' j_2)$. This illustrates the steps whereby sums of products of C-coefficients are evaluated. The C-coefficients disappear in pairs by the orthogonality relation; for each such pair one W-coefficient arises.

The last symmetry relations in (6.10a) and (6.10b) transform the Racah coefficient into a somewhat more orderly form, and the matrix element of $\mathbf{T}_L(1) \cdot \mathbf{T}_L(2)$ is finally written as

$$(j'm'j_1'j_2' | \mathbf{T}_L(1) \cdot \mathbf{T}_L(2) | j m j_1 j_2) = \delta_{m'm} \, \delta_{j'j} (-)^{j_1+j_2-j} \, W(j_1 j_2 j_1' j_2'; jL)$$
$$\times [(2j_1'+1)(2j_2'+1)]^{\frac{1}{2}} (j_1' \| T_L(1) \| j_1)(j_2' \| T_L(2) \| j_2) \quad (6.21)$$

If we compare this result with (6.19), where the same matrix element is expressed in terms of the reduced matrix element of the contraction

$\mathbf{T}_L(1)\cdot\mathbf{T}_L(2)$ in the representation that diagonalizes the total angular momentum, we find that

$$(jj_1'j_2' \parallel \mathbf{T}_L(1)\cdot\mathbf{T}_L(2) \parallel jj_1j_2) = (-)^{j_1'+j_2-j}[(2j_1'+1)(2j_2'+1)]^{\frac{1}{2}}$$

$$\times W(j_1j_2j_1'j_2'; jL)(j_1' \parallel T_L(1) \parallel j_1)(j_2' \parallel T_L(2) \parallel j_2) \quad (6.22)$$

Thus the Racah coefficients relate the reduced matrix elements of $T_L(1)$ in the j_1m_1 representation and $T_L(2)$ in the j_2m_2 representation to the reduced matrix of the contraction $\mathbf{T}_L(1)\cdot\mathbf{T}_L(2)$ in the coupled representation which diagonalizes the total angular momentum.

Another matrix element of interest is that of the tensor $T_L(1)$, which operates on only one part of a system, in the coupled representation in which the square and z-component of the total angular momentum are diagonal. An example is considered in Appendix I. As in the preceding example, the total angular momentum is the sum of the angular momenta of the two parts of the system: $\mathbf{j} = \mathbf{j}_1 + \mathbf{j}_2$. The matrix element is then

$$(j'm'j_1'j_2' \mid T_L(1) \mid jmj_1j_2) = \sum_{m_1'}\sum_{m_1} C(j_1'j_2'j'; m_1', m'-m_1')$$

$$\times C(j_1j_2j; m_1, m-m_1)(j_1'm_1' \mid T_L(1) \mid j_1m_1) \, \delta_{m'-m_1',m-m_1} \, \delta_{j_2'j_2} \quad (6.23)$$

Using the Wigner–Eckart theorem (5.14), the matrix element of the tensor in space 1 is

$$(j_1'm_1' \mid T_{LM}(1) \mid j_1m_1) = \delta_{m_1',m_1+M} \, C(j_1Lj_1'; m_1M)(j_1' \parallel T_L(1) \parallel j_1)$$

Then the sum over m_1' in (6.23) can be eliminated,

$$(j'm'j_1'j_2' \mid T_{LM}(1) \mid jmj_1j_2) = \delta_{m',m+M} \, \delta_{j_2'j_2}(j_1' \parallel T_L(1) \parallel j_1)$$

$$\times \sum_{m_1} C(j_1j_2j; m_1, m-m_1) \, C(j_1Lj_1'; m_1M) \, C(j_1'j_2'j'; m_1+M, m-m_1)$$

$$(6.23a)$$

If the symmetry relation (3.16c) is used on the third C-coefficient of this sum, the last two coefficients become

$$(-)^{j_1-m_1-M}\left[\frac{2j'+1}{2j_2+1}\right]^{\frac{1}{2}} C(j_1Lj_1'; m_1M) \, C(j_1'j'j_2; m_1+M, -m-M)$$

which means that (6.4b) can now be used. The sum over m_1 is thus

$$(2j'+1)(2j_2+1)^{-\frac{1}{2}} \sum_{m_1} (-)^{j_1-m_1-M} \, C(j_1j_2j; m_1, m-m_1)$$

$$\times \sum_f (2f+1)^{\frac{1}{2}} W(j\,L\,j\,j'; j'f) \, C(Lj'f; M, -m-M) \, C(j_1fj_2; m_1, -m)$$

The second C-coefficients can be removed from the sum over m_1, which is

$$\sum_{m_1} (-)^{j_1'-m_1-M} C(j_1 j_2 j; m_1, m-m_1) C(j_1 f j_2; m_1, -m)$$

$$= (-)^{j_1'-j_1-M} \left[\frac{2j_2 + 1}{2f + 1}\right]^{\frac{1}{2}} \sum_{m_1} C(j_1 j_2 j; m_1, m-m_1) C(j_1 j_2 f; m_1, m-m_1)$$

$$= (-)^{j_1'-j_1-M} \left[\frac{2j_2 + 1}{2j + 1}\right]^{\frac{1}{2}} \delta_{fj}$$

where the symmetry relation (3.16c) was used to get the second line, and the orthonormality property (3.7) was used to get the third line. When these results are collected, the original matrix element becomes

$$(j'm'j_1'j_2' | T_{LM}(1) | j m j_1 j_2) = \delta'_{j_2 j_2} \delta_{m',m+M} C(Lj'j; M, -m-M)$$

$$\times (-)^{j_1-j_1-M}[(2j_1' + 1)(2j' + 1)]^{\frac{1}{2}} W(j_1 L j_2 j'; j_1'j)(j_1' \| T_L(1) \| j_1)$$

If we apply symmetry relations (3.17b) and (3.16a) to the one remaining Clebsch–Gordan coefficient, and symmetry relations (6.10a) and (6.10b) to the Racah coefficient, we obtain

$$(j'm'j_1'j_2' | T_{LM}(1) | j m j_1 j_2) = C(jLj'; mMm') (-)^{j_2+L-j_1-j}$$

$$\times \delta_{j_2'j_2}[(2j_1' + 1)(2j + 1)]^{\frac{1}{2}} W(j_1 j j_1'j'; j_2 L)(j_1' \| T_L(1) \| j_1) \quad (6.24)$$

This is an expression of the Wigner–Eckart theorem (5.14) and implies the following relation between the reduced matrix element of $T_L(1)$ in the uncoupled representation and its reduced matrix element in the coupled representation:

$$(j'j_1'j_2' \| T_L \| j j_1 j_2) = \delta_{j_2'j_2}(-)^{j_2+L-j_1-j}[(2j_1' + 1)(2j + 1)]^{\frac{1}{2}}$$

$$\times W(j_1 j j_1'j'; j_2 L)(j_1' \| T_L(1) \| j_1) \quad (6.25)$$

As a further example of this type of result, we cite the case of matrix elements for beta emission, say, for two-particle configurations expressed in terms of those for single-particle configurations.[12] As far as the present discussion is concerned, the essential point is that (6.25) applies wherever the matrix elements of an operator expressed as a sum of single-particle operators are evaluated in the compound space of two-particle configurations. The factor multiplying the reduced matrix element in (6.25), which is just $(-)^{j_1-j_1}R_{jj'}$, may be thought of as the effect of the additional coupling $j_1 + j_2 = j$. That the result (6.25) is almost trivial is seen when we recall that we could couple j_1 and j_2 to obtain j

$$j_1 + j_2 = j \quad (6.26a)$$

[12] M. E. Rose and R. Osborne, *Phys. Rev.* **93**, 1326 (1954).

and then couple $\mathbf{j}$ to $\mathbf{L}$ to get $\mathbf{j}'$,

$$\mathbf{j} + \mathbf{L} = \mathbf{j}' \qquad (6.26b)$$

Alternatively, we could have

$$\mathbf{j}_1 + \mathbf{L} = \mathbf{j}_1' \qquad (6.27a)$$

and

$$\mathbf{j}_1' + \mathbf{j}_2 = \mathbf{j}' \qquad (6.27b)$$

Then the unitary matrix effecting the recoupling is just $R_{jj_1'}$.

Our discussion of the Racah and related coefficients can only begin to suggest their extremely widespread application. Some of the diverse fields covered are atomic and molecular spectra, directional correlation of nuclear radiations, angular distributions and polarization in nuclear reactions, and nuclear structure calculations. The common element in all these considerations is the fact that states of sharp angular momentum are involved. In transitions between such states, for example, radiation emitted or absorbed has sharp angular momentum or can be thought of as a mixture of a number of definite angular momenta. The selection rule expressed by the Wigner–Eckart theorem then serves to select particular values L of these angular momenta consistent with the conservation relation. Parity requirements may eliminate every other possible L value. For details, one may refer to the literature in these subjects. For instance, the theory of angular momentum has been of particular importance in the theory of angular correlation of nuclear radiations. A brief description of the theory is given in section 33. Nuclear reactions have been treated by Blatt and Biedenharn.[13]

25. THE GRADIENT FORMULA

For applications to follow it is important to evaluate the gradient of the irreducible tensor $\Phi(r)\, Y_{lm}$ where Φ is an arbitrary (differentiable) function of the scalar distance r. As an incidental point, the derivation of this gradient formula provides another simple application of the Racah algebra.

We consider $\mathbf{B} = \nabla \Phi(r)\, Y_{lm}$ and observe that the gradient operator itself is a first-rank tensor. This is verified by noting that (5.12) would now become

$$[L_\mu, \nabla_m] = (-)^\mu \sqrt{2}\; C(111; m+\mu, -\mu)\, \nabla_{m+\mu} \qquad (5.12a)$$

where L_μ are, as usual, the components of the orbital angular momentum in the spherical basis and ∇_m are the components of the gradient operator in the same basis. With the aid of Table I.2 it is seen that (5.12a)

[13] J. M. Blatt and L. C. Biedenharn, *Revs. Mod. Phys.* **24**, 258 (1952).

is indeed valid. It now follows that $\mathbf{B}$ can be expressed in terms of the irreducible spherical tensors $\mathbf{T}_{JLM}$ introduced in section 21 [see equation (5.57)] and that we must have the triangular relation $\Delta(Jl1)$. Moreover the parity of $\mathbf{B}$ is $(-)^{l+1}$. Therefore, the parity of $\mathbf{T}_{JLM}$ is $\pi_L = (-)^L = (-)^{l+1}$. Since we also have $\Delta(JL1)$, it follows that L can be equal to $l+1$ and $l-1$ only.

We can also use the commutation rules (5.12) to show that $\mathbf{B}$ is an irreducible tensor of rank l so that $J = l$ only. However, this result arises automatically in the process of evaluating the coefficients in the tensor expansion of $\mathbf{B}$. Thus,

$$\mathbf{B} = \sum_{JLM} C_{JLM}\, \mathbf{T}_{JLM} \tag{6.28}$$

and by virtue of (5.60)

$$C_{JLM} = \int d\Omega\, \mathbf{T}^*_{JLM}\cdot\mathbf{B} \tag{6.29}$$

We write the gradient operator as a sum of two parts

$$\nabla = \mathbf{r}(\mathbf{r}\cdot\nabla) - \mathbf{r}\times(\mathbf{r}\times\nabla)$$

$$= \mathbf{r}\frac{\partial}{\partial r} - \frac{i}{r}(\mathbf{r}\times\mathbf{L}) \tag{6.30}$$

and the corresponding decomposition of the coefficients is written

$$C_{JLM} = C'_{JLM} + C''_{JLM} \tag{6.31}$$

where

$$C'_{JLM} = \frac{\partial\Phi}{\partial r}\int d\Omega\, \mathbf{T}^*_{JLM}\cdot\mathbf{r}\, Y_{lm} \tag{6.31a}$$

and

$$C''_{JLM} = -\frac{i\Phi}{r}\int d\Omega\, \mathbf{T}^*_{JLM}\cdot\mathbf{r}\times\mathbf{L}\, Y_{lm} \tag{6.31b}$$

To evaluate C'_{JLM} we use

$$\mathbf{r}\, Y_{lm} = \left(\frac{4\pi}{3}\right)^{\frac{1}{2}}\sum_{\mu}(-)^{\mu}\, Y_{1\mu}\, Y_{lm}\, \boldsymbol{\xi}_{-\mu}$$

$$= \sum_{\lambda\mu}(-)^{\mu}\, \boldsymbol{\xi}_{-\mu}\left(\frac{2l+1}{2\lambda+1}\right)^{\frac{1}{2}} C(l1\lambda;00)\, C(l1\lambda;m\mu)\, Y_{\lambda,m+\mu}$$

by the coupling relation (4.32). With the use of (3.17a) and (3.16a) this becomes

$$\mathbf{r}\, Y_{lm} = -\sum_{\lambda} C(l1\lambda;00)\sum_{\mu} C(\lambda 1l;m+\mu,-\mu)\, Y_{\lambda,m+\mu}\, \boldsymbol{\xi}_{-\mu}$$

and we have used the fact that $\lambda + 1 + l$ must be even. From the definition (5.57) this is

$$\mathbf{r}\, Y_{lm} = - \sum_{\lambda} C(l1\lambda; 00)\, \mathbf{T}_{l\lambda m} \tag{6.32}$$

Of course, $\lambda = l + 1$ and $l - 1$, as was pointed out above. This gives

$$C'_{JLM} = - \delta_{Jl}\, \delta_{Mm}\, \frac{\partial \Phi}{\partial r}\, C(l1L; 00) \tag{6.33}$$

with $L = l \pm 1$.

The second part of the coefficient C_{JLM} is more complicated. To express $\mathbf{r} \times \mathbf{L}\, Y_{lm}$ in terms of the spherical tensors, we use

$$\mathbf{L}\, Y_{lm} = \sum_{\mu} (-)^{\mu}\, L_{\mu}\, Y_{lm}\, \mathbf{\xi}_{-\mu}$$

$$= [l(l + 1)]^{\frac{1}{2}} \sum_{\mu} C(l1l; m+\mu, -\mu)\, Y_{l,m+\mu}\, \mathbf{\xi}_{-\mu} \tag{6.34}$$

by (5.13). Again this would be written in the form

$$\mathbf{L}\, Y_{lm} = [l(l + 1)]^{\frac{1}{2}}\, \mathbf{T}_{llm} \tag{6.34'}$$

and this result will be useful later. However, for our present purpose it is not especially convenient. We now need

$$\mathbf{r} \times \mathbf{L}\, Y_{lm} = \left(\frac{4\pi}{3}\right)^{\frac{1}{2}} \sum_{\mu'} (-)^{\mu'}\, Y_{1\mu'}\, \mathbf{\xi}_{-\mu'} \times \mathbf{L}\, Y_{lm}$$

$$= \left[\frac{4\pi}{3}\, l(l + 1)\right]^{\frac{1}{2}} \sum_{\mu\mu'} (-)^{\mu'}\, Y_{1\mu'}\, Y_{l,m+\mu}\, C(l1l; m+\mu, -\mu)$$

$$\times (\mathbf{\xi}_{-\mu'} \times \mathbf{\xi}_{-\mu}) \tag{6.35}$$

A table of all possible cross-products of the spherical basis vectors can be made. The results are then seen to be expressible in the form

$$\mathbf{\xi}_{\mu'} \times \mathbf{\xi}_{\mu} = i\, S(\mu' - \mu)\, \mathbf{\xi}_{\mu'+\mu}$$

where $S(\mu' - \mu)$ is the sign of $\mu' - \mu$ and is zero if $\mu' - \mu = 0$. This form is simple enough, but, like all other quantities depending on projection quantum numbers, it must be expressed in terms of C-coefficients. Then, from Table I.2, it is seen that

$$\mathbf{\xi}_{\mu'} \times \mathbf{\xi}_{\mu} = i\sqrt{2}\, C(111; \mu'\mu)\, \mathbf{\xi}_{\mu'+\mu} \tag{6.36}$$

This result is now inserted in (6.35), and (4.32) is used to express the product of spherical harmonics as a sum of harmonics. After introduction of the summation index $\nu = \mu + \mu'$, the result is

$$i(\mathbf{r} \times \mathbf{L}) \, Y_{lm} = -[2l(l + 1)(2l + 1)]^{\frac{1}{2}} \sum_{\nu\lambda} (-)^{\nu}(2\lambda + 1)^{-\frac{1}{2}}$$

$$\times C(l1\lambda; 00) \, Y_{\lambda,m+\nu} \, \boldsymbol{\xi}_{-\nu} \, S_{\nu} \quad (6.37)$$

where S_{ν} is a sum of products of three C-coefficients:

$$S_{\nu} = \sum_{\mu} (-)^{\mu} C(l1\lambda; m{+}\mu, \nu{-}\mu) \, C(l1l; m{+}\mu, -\mu) \, C(111; -\nu{+}\mu, -\mu)$$

This sum is just of the general type shown in (6.6b). Its evaluation in terms of a Racah coefficient is straightforward, and the result is

$$S_{\nu} = [3(2l + 1)]^{\frac{1}{2}} C(l1\lambda; m\nu) \, W(l1\lambda 1; l1)$$

$$= (-)^{1+\nu}[3(2\lambda + 1)]^{\frac{1}{2}} C(\lambda 1 l; m{+}\nu, -\nu) \, W(l\lambda 11; 1l) \quad (6.38)$$

by (3.17a) and the fact that $\lambda + 1 + l$ is an even integer. We have also used (6.10a) to write the arguments of the Racah coefficient in a different order.

With the result (6.38) we find

$$i(\mathbf{r} \times \mathbf{L}) \, Y_{lm} = [6l(l + 1)(2l + 1)]^{\frac{1}{2}} \sum_{\lambda} C(l1\lambda; 00) \, W(l\lambda 11; 1l) \, \mathbf{T}_{l\lambda m}$$

$$(6.39)$$

where again (5.57) has been used. It is interesting to note that both radial and tangential parts of the vector $\nabla\Phi \, Y_{lm}$ are spherical tensors of rank l. Also both parts have the same projection quantum number as does the spherical harmonic. We are now able to write

$$C''_{JLM} = -[6l(l + 1)(2l + 1)]^{\frac{1}{2}} \frac{\Phi}{r} \delta_{Jl} \, \delta_{Mm} \, C(l1L; 00) \, W(lL11; 1l) \quad (6.40)$$

where again $L = l \pm 1$. The final result for the expansion coefficients is now

$$C_{JLM} = -\delta_{Jl} \, \delta_{Mm} \, C(l1L; 00) \left\{ \frac{\partial\Phi}{\partial r} + [6l(l + 1)(2l + 1)]^{\frac{1}{2}} \, W(lL11; 1l) \frac{\Phi}{r} \right\}$$

$$(6.41)$$

and $L = l \pm 1$ only, as the parity C-coefficient requires.

From Table I.4 the values of the Racah coefficients are

$$W(l \, l{+}1 \, 11; 1l) = -[l/6(l + 1)(2l + 1)]^{\frac{1}{2}}$$

$$W(l \, l{-}1 \, 11; 1l) = [(l + 1)/6l(2l + 1)]^{\frac{1}{2}}$$

Inserting these results, we arrive at the final form for the gradient formula

$$\mathbf{B} \equiv \nabla\Phi(r)\, Y_{lm} = -\left(\frac{l+1}{2l+1}\right)^{\frac{1}{2}} \left(\frac{d\Phi}{dr} - l\frac{\Phi}{r}\right) \mathbf{T}_{l,l+1,m}$$

$$+ \left(\frac{l}{2l+1}\right)^{\frac{1}{2}} \left(\frac{d\Phi}{dr} + \frac{l+1}{r}\Phi\right) \mathbf{T}_{l,l-1,m} \quad (6.42)$$

This result will be applied in sections 27 to 29. It may be useful in other connections to remember that the radial and tangential parts of **B** are clearly separated in (6.42) insofar as the former is simply the terms with $d\Phi/dr$ and the latter the terms in Φ/r. As a check we can use

$$\mathbf{r} \cdot \mathbf{T}_{JLM} = -C(J1L; 00)\, Y_{JM} \quad (6.43)$$

for the radial part of the tensors. This is obtained directly from (5.57) and (3.7) and gives an alternative derivation of (6.33). This result also gives

$$\mathbf{r}(\mathbf{r} \cdot \mathbf{T}_{JLM}) = \sum_{L'} C(J1L; 00)\, C(J1L'; 00)\, \mathbf{T}_{JL'M}$$

from which the tangential part of the tensor is easily obtained.

PART B

Applications

VII. THE ELECTROMAGNETIC FIELD

26. THE MAXWELL EQUATIONS

In discussing the electromagnetic interactions of a system whose quantum states have definite angular momentum and parity, it is appropriate to describe the electromagnetic field in spherical rather than Cartesian coordinates. In this way one is led to the multipole fields, those solutions of the Maxwell equations that are simultaneously eigenfunctions of the angular momentum of the radiation field, the component thereof on some arbitrarily selected quantization axis and also of the parity operator. Such multipole fields have been discussed by many authors.[1] In this section we shall outline and discuss a few of the pertinent properties of these fields.

Maxwell's electromagnetic field equations are

$$\nabla \times \mathbf{H}' = \frac{4\pi}{c}\mathbf{j}' + \frac{1}{c}\frac{\partial \mathbf{E}'}{\partial t} \qquad (7.1a)$$

$$\nabla \times \mathbf{E}' = -\frac{1}{c}\frac{\partial \mathbf{H}'}{\partial t} \qquad (7.1b)$$

$$\nabla \cdot \mathbf{H}' = 0 \qquad (7.1c)$$

$$\nabla \cdot \mathbf{E}' = 4\pi\rho' \qquad (7.1d)$$

The vacuum velocity of light is designated by c. Primes indicate the real observed quantities. For monochromatic radiation of frequency ω we introduce complex, time-independent, amplitudes.[2] For the field strengths we write

$$\mathbf{H}'_{\omega}(\mathbf{r}, t) = \mathbf{H}_{\omega}(\mathbf{r})\, e^{-i\omega t} + \mathbf{H}^*_{\omega}(\mathbf{r})\, e^{i\omega t} \qquad (7.2a)$$

$$\mathbf{E}'_{\omega}(\mathbf{r}, t) = \mathbf{E}_{\omega}(\mathbf{r})\, e^{-i\omega t} + \mathbf{E}^*_{\omega}(\mathbf{r})\, e^{i\omega t} \qquad (7.2b)$$

[1] The subject matter of this section is treated more extensively in M. E. Rose, *Multipole Fields*, John Wiley & Sons, New York, 1955. For additional literature on this subject consult the bibliography given therein.

[2] We are following the notation of section 2 of reference 1, except that primes are used instead of bars to denote time-dependent quantities.

and for the current and charge densities we introduce

$$\mathbf{j}'_\omega(\mathbf{r}, t) = \mathbf{j}_\omega(\mathbf{r})\, e^{-i\omega t} + \mathbf{j}^*_\omega(\mathbf{r})\, e^{i\omega t} \qquad (7.2c)$$

$$\rho_\omega(\mathbf{r}, t) = \rho_\omega(\mathbf{r})\, e^{-i\omega t} + \rho^*_\omega(\mathbf{r})\, e^{i\omega t} \qquad (7.2d)$$

It will not be necessary to carry the subscript ω in the following discussion, and it is henceforth dropped. With introduction of the wave number $k = \omega/c$, Maxwell's equations become

$$\nabla \times \mathbf{H} = \frac{4\pi}{c}\, \mathbf{j} - ik\mathbf{E} \qquad (7.3a)$$

$$\nabla \times \mathbf{E} = ik\mathbf{H} \qquad (7.3b)$$

$$\nabla \cdot \mathbf{H} = 0 \qquad (7.3c)$$

$$\nabla \cdot \mathbf{E} = 4\pi\rho \qquad (7.3d)$$

In the usual way, a vector potential $\mathbf{A}'$ may be introduced, as implied by (7.1c),

$$\mathbf{H}' = \nabla \times \mathbf{A}' \qquad (7.4a)$$

Corresponding to (7.2), we have

$$\mathbf{A}'(\mathbf{r}, t) = \mathbf{A}(\mathbf{r})\, e^{-i\omega t} + \mathbf{A}^*(\mathbf{r})\, e^{i\omega t}$$

and

$$\mathbf{H} = \nabla \times \mathbf{A}$$

With this result, (7.1b) permits the use of a scalar potential ϕ', such that

$$\mathbf{E}' = -\nabla\phi' - \frac{1}{c}\frac{\partial \mathbf{A}'}{\partial t} \qquad (7.4b)$$

Again, the time-independent potential ϕ is introduced by

$$\phi'(\mathbf{r}, t) = \phi(\mathbf{r})\, e^{-i\omega t} + \phi^*(\mathbf{r})\, e^{i\omega t}$$

so that we have

$$\mathbf{E} = -\nabla\phi + ik\mathbf{A}$$

The electric and magnetic fields are therefore determined by the vector and scalar potentials. On the other hand, the fields do not completely determine the potentials, since any scalar function S' can be introduced such that $\mathbf{A}'_r$ and ϕ'_r, defined by

$$\mathbf{A}'_r = \mathbf{A}' + \nabla S' \qquad (7.5a)$$

$$\phi'_r = \phi' - \frac{1}{c}\frac{\partial S'}{\partial t} \qquad (7.5b)$$

are satisfactory vector and scalar potentials if $\mathbf{A}'$ and ϕ' are. That is, Maxwell's equations are invariant to the introduction of the function S', which is called a gauge transformation. The reason for this is that the fields evaluated from (7.4a) and (7.4b) are not changed. Substitution of (7.4a) and (7.4b) into (7.3a) and (7.3d) yield the following equations for $\mathbf{A}'$ and ϕ':

$$\nabla^2\mathbf{A}' - \frac{1}{c^2}\frac{\partial^2\mathbf{A}'}{\partial t^2} = -\frac{4\pi}{c}\mathbf{j}' + \nabla\left(\nabla\cdot\mathbf{A}' + \frac{1}{c}\frac{\partial\phi'}{\partial t}\right) \qquad (7.6a)$$

$$\nabla^2\phi' = -4\pi\rho' - \frac{1}{c}\frac{\partial}{\partial t}(\nabla\cdot\mathbf{A}') \qquad (7.6b)$$

Probably the most familiar gauge transformation is that which carries arbitrary vector and scalar potentials $\mathbf{A}_0'$, ϕ_0' into ones satisfying the Lorentz condition.

$$\nabla\cdot\mathbf{A}' + \frac{1}{c}\frac{\partial\phi'}{\partial t} = 0 \qquad (7.7)$$

According to (7.5), this gauge transformation is a solution of the inhomogeneous wave equation

$$\nabla^2 S' - \frac{1}{c^2}\frac{\partial^2 S'}{\partial t^2} = -\left(\nabla\cdot\mathbf{A}_0' + \frac{1}{c}\frac{\partial\phi_0'}{\partial t}\right) \qquad (7.8a)$$

Similarly, the Lorentz condition is invariant under gauge transformations which are solutions of the homogeneous wave equation. The advantage of the Lorentz condition is that the partial differential equations for the scalar and vector potentials are the simple and familiar wave equations

$$\nabla^2\phi' - \frac{1}{c^2}\frac{\partial^2\phi'}{\partial t^2} = -4\pi\rho' \qquad (7.9a)$$

$$\nabla^2\mathbf{A}' - \frac{1}{c^2}\frac{\partial^2\mathbf{A}'}{\partial t^2} = -\frac{4\pi}{c}\mathbf{j}' \qquad (7.9b)$$

so that the equations for ϕ' and $\mathbf{A}'$ are decoupled. When $\mathbf{A}_0'$, ϕ_0' satisfy the Lorentz condition, the gauge function S' is a solution of the (vacuum) homogeneous wave equation,

$$\nabla^2 S' - \frac{1}{c^2}\frac{\partial^2 S'}{\partial t^2} = 0 \qquad (7.8b)$$

It is convenient to have the results of the preceding paragraph in time-independent form. Thus the electromagnetic field amplitudes are determined from the vector and scalar potential amplitudes:

$$\mathbf{H} = \nabla\times\mathbf{A}, \qquad \mathbf{E} = -\nabla\phi + ik\mathbf{A} \qquad (7.10)$$

A gauge transformation S with the corresponding transformation equations

$$\mathbf{A}_\tau = \mathbf{A} + \nabla S \qquad (7.11)$$

$$\phi_\tau = \phi + ikS \qquad (7.12)$$

does not alter the field strengths as given by (7.10), and therefore it leaves unchanged the form of Maxwell's equations (7.1).

In the Lorentz gauge

$$\nabla \cdot \mathbf{A} - ik\phi = 0 \qquad (7.13)$$

the vector and scalar potentials satisfy the inhomogeneous wave equations

$$\nabla^2 \mathbf{A} + k^2 \mathbf{A} = -\frac{4\pi}{c}\mathbf{j} \qquad (7.14a)$$

$$\nabla^2 \phi + k^2 \phi = -4\pi\rho \qquad (7.14b)$$

Gauge transformations which satisfy the homogeneous wave equation

$$\nabla^2 S + k^2 S = 0 \qquad (7.15)$$

clearly preserve the Lorentz condition. In source-free regions, $\rho = 0$, $\mathbf{j} = 0$, over some part of space, this is the equation satisfied by the scalar potential. Thus $S = i\phi/k$ is a gauge transformation which preserves the Lorentz condition. Furthermore it transforms the scalar potential to zero so that the fields are determined by the vector potential. This is the so-called solenoidal gauge. The scalar potential is zero

$$\phi = 0 \qquad (7.16a)$$

and the Lorentz condition is

$$\nabla \cdot \mathbf{A} = 0 \qquad (7.16b)$$

which explains the nomenclature.

Assuming that ρ and $\mathbf{j}$ are confined to a limited portion of space, the vector and scalar potentials are solutions of the homogeneous vector and scalar Helmholtz equations in the outside region. Thus

$$\nabla^2 \mathbf{A} + k^2 \mathbf{A} = 0 \qquad (7.17a)$$

$$\nabla^2 \phi + k^2 \phi = 0 \qquad (7.17b)$$

The solutions to the inhomogeneous wave equations (7.14a) and (7.14b) can also be expressed in terms of these by using Green's function techniques.[3]

---

[3] For example, see J. Blatt and V. Weisskopf, *Theoretical Nuclear Physics*, Appendix B, John Wiley & Sons, New York, 1952.

An important property of the solutions to (7.17a) and (7.17b) is their parity, i.e., their behavior under inversion of the coordinates: $\mathbf{r} \rightarrow -\mathbf{r}$. Since the operator $(\nabla^2 + k^2)$ is an even operator, the potentials $\mathbf{A}(-\mathbf{r})$, $\phi(-\mathbf{r})$ are also solutions of the wave equations. Therefore, the linear combinations $\mathbf{A}(\mathbf{r}) \pm \mathbf{A}(-\mathbf{r})$, $\phi(\mathbf{r}) \pm \phi(-\mathbf{r})$ are solutions with even (+ sign) or odd (− sign) parity. Of course, an even function is unaltered; an odd function changes sign under inversion. From (7.10) we see that the electric field has the same parity as the vector potential and the magnetic field has parity opposite to that of the vector potential. Thus, if $\mathbf{A}$ has a definite parity, the electric field and the magnetic field have opposite parities. This was also clear from the original Maxwell equations (7.1b).

If we consider the source-free Maxwell equations

$$\nabla \times \mathbf{H}' = \frac{1}{c}\frac{\partial \mathbf{E}'}{\partial t}, \qquad \nabla \times \mathbf{E}' = -\frac{1}{c}\frac{\partial \mathbf{H}'}{\partial t}$$

$$\nabla \cdot \mathbf{H}' = 0, \qquad\qquad \nabla \cdot \mathbf{E}' = 0 \qquad\qquad (7.18)$$

we notice that $\mathbf{E}'_d = \mp \mathbf{H}'$ and $\mathbf{H}'_d = \pm \mathbf{E}'$ are solutions if $\mathbf{E}'$ and $\mathbf{H}'$ are solutions, where either choice of sign may be used. The fields $\mathbf{E}'_d$, $\mathbf{H}'_d$ are said to be dual to the fields $\mathbf{E}'$, $\mathbf{H}'$. The important property of the dual fields is that they have the same energy density $(1/8\pi)(\mathbf{E}'^2 + \mathbf{H}'^2)$, Poynting vector, $(c/4\pi)\mathbf{E}' \times \mathbf{H}'$, and angular momentum density $(4\pi c)^{-1}\mathbf{r} \times (\mathbf{E}' \times \mathbf{H}')$ as the original fields.

27. THE MULTIPOLE FIELDS

With these preliminary remarks as a background, we can now proceed to the construction of the multipole electromagnetic fields. A 2^L pole multipole field is a solution of the free-space Maxwell equations which is an eigenfunction of the square and one component of the total angular momentum, the eigenvalue of the former operator being $L(L+1)$, and of the parity. Thus the potentials in a 2^L pole multipole field must be irreducible tensors of rank L with definite parity. The parity is determined by the parity of the magnetic field $\mathbf{H}$. A multipole field is a 2^L pole electric field if the parity of the magnetic field is $(-)^L$. It is a 2^L pole magnetic field if the parity of the magnetic field is $(-)^{L+1}$. Then, as will be clear from the discussion to follow, the electric and magnetic multipole fields for given L are dual fields. Since the dual property refers to field strengths and not to potentials, there can be no question of differences due to gauge transformations.

To construct the multipole fields we first consider the simple case of the scalar Helmholtz equation. The solution with the required angular momentum and parity properties is

$$\psi_{LM} = \zeta_L(kr)\, Y_{LM}(\theta\varphi) \tag{7.19}$$

where ζ_L, the radial function, satisfies

$$\frac{d^2\zeta_L}{dx^2} + \frac{2}{x}\frac{d\zeta_L}{dx} + \left(1 - \frac{L(L+1)}{x^2}\right)\zeta_L = 0 \tag{7.20}$$

where $x = kr$. There are, of course, two linearly independent solutions of (7.20). That one which is regular at the origin is $j_L(x)$ (with $L \geq 0$), the spherical Bessel function

$$j_L(x) = \left(\frac{\pi}{2x}\right)^{\frac{1}{2}} J_{L+\frac{1}{2}}(x) \tag{7.20a}$$

A solution which is not regular at the origin is the spherical Neumann function [4]

$$n_L(x) = (-)^{L+1}\left(\frac{\pi}{2x}\right)^{\frac{1}{2}} J_{-L-\frac{1}{2}}(x) \tag{7.20b}$$

In fact, for $x \to 0$

$$j_L(x) = \frac{x^L}{1\cdot3\cdot5\cdots(2L+1)} \tag{7.21a}$$

while

$$n_L(x) = -\frac{1\cdot3\cdot5\cdots(2L-1)}{x^{L+1}} \tag{7.21b}$$

At infinity both $j_L(x)$ and $n_L(x)$ are standing waves. Thus

$$j_L(x) \to \frac{1}{x}\sin\left(x - \frac{L\pi}{2}\right) \tag{7.22a}$$

$$n_L(x) \to -\frac{1}{x}\cos\left(x - \frac{L\pi}{2}\right) \tag{7.22b}$$

This shows that the spherical Hankel function of the first kind defined by

$$h_L^{(1)}(x) = j_L(x) + i\, n_L(x) \tag{7.23a}$$

is an outgoing wave at infinity:

$$h_L^{(1)}(x) \to \frac{(-i)^{L+1}}{x} e^{ix} \tag{7.23b}$$

[4] E. Jahnke and F. Emde, *Tables of Functions*, p. 131, Dover Press, New York, 1943.

We therefore distinguish between standing-wave solutions

$$U_{LM} = j_L(kr)\, Y_{LM}(\theta, \varphi) \qquad\qquad (7.24a)$$

and outgoing-wave solutions

$$V_{LM} = h_L^{(1)}(kr)\, Y_{LM}(\theta, \varphi) \qquad\qquad (7.24b)$$

For both of these the transformation properties under rotation depend only on the spherical harmonics so that U_{LM} and V_{LM} are both irreducible tensors of rank L. Similarly the parity operation affects only the angular part of the fields and consists of replacing θ by $\pi - \theta$ and φ by $\pi + \varphi$. Referring to equation (III.20) of Appendix III we have that

$$Y_{LM}(\pi - \theta, \pi + \varphi) = (-)^L\, Y_{LM}(\theta, \varphi) \qquad\qquad (7.25)$$

so that the U_{LM} and V_{LM} have the parity $(-)^L$.

The solutions of the homogeneous vector Helmholtz equation can now be constructed in terms of the solutions introduced above for the scalar equation. At the same time we wish to impose the condition that the vector potential be an eigenfunction of $\mathbf{J}^2$, J_z, and the parity operator. Of course $\mathbf{J}$ is as defined in section 21. That this will also be true for the scalar potential ϕ as well as for the electric and magnetic fields will be apparent from results given below.

The vector potential for the magnetic multipole field with angular momentum L must have parity $(-)^L$ by the definition given above.[5] Then

$$\mathbf{A}_{LM}(m) = C_L(m)\, \zeta_L\, \mathbf{T}_{LLM} \qquad\qquad (7.26)$$

has all the required properties. Here $C_L(m)$ is an arbitrary normalization factor. That (7.26) is an eigenfunction of $\mathbf{J}^2$ and J_z with eigenvalues $L(L + 1)$ and M follows from the discussion of section 21. The parity parameter λ in $\mathbf{T}_{L\lambda M}$ can be only L, $L \pm 1$ and by definition only L is permitted for the magnetic multipole. Since $\mathbf{T}_{L\lambda M}$ is a linear combination of $Y_{\lambda M'}$ it follows that $\zeta_\lambda\, \mathbf{T}_{L\lambda M}$ is a solution of the Helmholtz equation. The physical boundary conditions dictate the form of the radial function.

For the electric multipole of the same total angular momentum we must choose the tensors $\mathbf{T}_{L\lambda M}$ with $\lambda = L \pm 1$. Both of these have the correct parity and angular momentum properties. When they are multiplied with ζ_λ, they are proper solutions of the wave equation. Hence

$$\mathbf{A}_{LM}(e) = C_{L+1}(e)\, \zeta_{L+1}\, \mathbf{T}_{L,L+1,M} + C_{L-1}(e)\, \zeta_{L-1}\, \mathbf{T}_{L,L-1,M} \qquad (7.27)$$

[5] In order to make our notation conform with the customary one we use $L(L + 1)$ for the eigenvalues of $\mathbf{J}^2$.

is the appropriate form for the vector potential of the electric multipole field. Here $C_{L+1}(e)$ and $C_{L-1}(e)$ are two constants which represent a normalization and an arbitrary gauge constant. This remark is clarified below.

To calculate the scalar potential we need to calculate $\nabla \cdot \mathbf{A}$ and then use (7.13). This calculation is carried out with the aid of the gradient formula. Thus,

$$\nabla \cdot \zeta_\lambda \, \mathbf{T}_{L\lambda M} = \sum_\mu (-)^\mu \, \nabla_\mu \zeta_\lambda \, \mathbf{T}_{L\lambda M} \cdot \boldsymbol{\xi}_{-\mu}$$

$$= \sum_\mu C(\lambda 1 L; \mu, M-\mu) \, \nabla_\mu \, Y_{\lambda, M-\mu} \, \zeta_\lambda$$

by virtue of (5.56) and (5.57). Applying (6.42), we find

$$\nabla \cdot \zeta_\lambda \, \mathbf{T}_{L\lambda M} = \left(\frac{\lambda+1}{2\lambda+3}\right)^{\frac{1}{2}} \delta_{L,\lambda+1} \, Y_{\lambda+1,M} \left(\frac{d\zeta_\lambda}{dr} - \frac{\lambda}{r}\zeta_\lambda\right)$$

$$- \left(\frac{\lambda}{2\lambda-1}\right)^{\frac{1}{2}} \delta_{L\,\lambda-1} \, Y_{\lambda-1,M} \left(\frac{d\zeta_\lambda}{dr} + \frac{\lambda+1}{r}\zeta_\lambda\right)$$

where we have used (3.7). If we take advantage of the properties of the spherical Bessel and Hankel functions, namely,

$$\frac{d\zeta_\lambda}{dx} = \frac{\lambda}{x}\zeta_\lambda - \zeta_{\lambda+1} = -\frac{\lambda+1}{x}\zeta_\lambda + \zeta_{\lambda-1}$$

with $x = kr$, we obtain [6]

$$\nabla \cdot \zeta_\lambda \, \mathbf{T}_{L\lambda M} = -\left[\left(\frac{L+1}{2L+1}\right)^{\frac{1}{2}} \delta_{\lambda,L+1} + \left(\frac{L}{2L+1}\right)^{\frac{1}{2}} \delta_{\lambda,L-1}\right] k\zeta_L \, Y_{LM}$$

$$\tag{7.28}$$

This gives

$$\phi_{LM}(m) = 0 \tag{7.29}$$

and

$$\phi_{LM}(e) = i\left[C_{L+1}\left(\frac{L+1}{2L+1}\right)^{\frac{1}{2}} + C_{L-1}\left(\frac{L}{2L+1}\right)^{\frac{1}{2}}\right] \zeta_L \, Y_{LM} \tag{7.30}$$

which is, of course, an irreducible tensor of rank L.

We make a gauge transformation with a (time-independent) gauge function

$$S = \frac{a}{k}\zeta_L \, Y_{LM} \tag{7.31}$$

---

[6] Formulae of this type have been given by E. L. Hill, *Am. J. Phys.* **22**, 211 (1954).

where a is an arbitrary constant. This gauge function is the only one that will not change the multipolarity properties of the field. We find that $(\phi_{LM})_\tau$ has the same form as (7.30) and $(\mathbf{A}_{LM})_\tau$ has the same form as (7.27). This is obvious from (7.11) and (6.42). In fact C_{L+1} is replaced by

$$C_{L+1} + a \left(\frac{L+1}{2L+1} \right)^{\frac{1}{2}} \tag{7.31a}$$

and C_{L-1} is replaced by

$$C_{L-1} + a \left(\frac{L}{2L+1} \right)^{\frac{1}{2}} \tag{7.31b}$$

This shows that the two constants in the electric multipole field are adjustable by a gauge transformation.

In particular, we can choose the solenoidal gauge which requires that

$$\frac{C_{L-1}}{C_{L+1}} = - \left(\frac{L+1}{L} \right)^{\frac{1}{2}} \tag{7.31c}$$

and then

$$\phi_{LM}(e) = 0 \tag{7.32a}$$

and, with $C_{L+1}^2 + C_{L-1}^2 = 1$, we can make the phase choice $C_{L+1} < 0$, so that

$$\mathbf{A}_{LM}(e) = - \left(\frac{L}{2L+1} \right)^{\frac{1}{2}} \zeta_{L+1} \, \mathbf{T}_{L,L+1,M} + \left(\frac{L+1}{2L+1} \right)^{\frac{1}{2}} \zeta_{L-1} \, \mathbf{T}_{L,L-1,M} \tag{7.32b}$$

With this result both magnetic and electric multipole fields are in the convenient solenoidal gauge. The field strengths can be calculated from these, as well as any other gauge. To make the fields dual we choose $C_L(m) = 1$. The results, which the reader may verify,[1] are

$$\mathbf{E}_{LM}(m) = -\mathbf{H}_{LM}(e) = -ik \, \zeta_L \, \mathbf{T}_{LLM} \tag{7.33}$$

$$\mathbf{H}_{LM}(m) = \mathbf{E}_{LM}(e) = ik \left[- \left(\frac{L}{2L+1} \right)^{\frac{1}{2}} \zeta_{L+1} \, \mathbf{T}_{L,L+1,M} \right.$$

$$\left. + \left(\frac{L+1}{2L+1} \right)^{\frac{1}{2}} \zeta_{L-1} \, \mathbf{T}_{L,L-1,M} \right] \tag{7.34}$$

These results bear out the statements made concerning the field strengths at the beginning of this section.

The multipole fields do not form a complete set since they represent only two linear combinations of three linearly independent tensors $\mathbf{T}_{L\lambda M}$

with $\lambda = L$ and $L \pm 1$. To form a complete set, any other linearly independent combination of two of these tensors will do. The customary combination is simply the longitudinal field whose vector potential is

$$\mathbf{A}_{LM}(l) = \frac{1}{k} \nabla \zeta_L Y_{LM} \qquad (7.35)$$

As already stated, this has the form of (7.27) but with C_{L+1} replaced by $[(L+1)/(2L+1)]^{\frac{1}{2}}$ and C_{L-1} by $[L/(2L+1)]^{\frac{1}{2}}$. The corresponding scalar potential is obviously $\phi_{LM}(l) = i\,\zeta_L\,Y_{LM}$. These longitudinal field potentials are important where radiation processes are completely forbidden. These are the cases where $L = 0$ is the only possibility because in that event all the field strengths, and hence the Poynting flux, vanish. Therefore in transitions between states of zero angular momentum the longitudinal field is all that contributes, and for other cases it can make a non-vanishing, though usually small, contribution to transition rates.[7]

As an application of some of the foregoing results, we consider the expansion of a plane wave into multipole fields. This is the analogue of the Rayleigh expansion but applicable to intrinsic spin one. It is important for many problems involving emission or absorption of photons. We consider a circularly polarized wave since there is no loss of generality thereby. The vector potential is

$$\mathbf{A} = \mathbf{u}_p\, e^{i\mathbf{k}\cdot\mathbf{r}}$$

where $\mathbf{u}_p$ is a polarization vector. If $\mathbf{u}_1$ and $\mathbf{u}_2$ are unit vectors which together with $\mathbf{f}$ form a right-handed coordinate system, we write

$$\mathbf{u}_p = \frac{1}{\sqrt{2}}\,(\mathbf{u}_1 + ip\mathbf{u}_2), \qquad p = \pm 1$$

and $p = 1$ corresponds to left circular polarization while $p = -1$ corresponds to right circular polarization.

The calculation is readily simplified by choosing the z-axis to coincide with $\mathbf{f}$ in direction. Then $\mathbf{u}_1 = \mathbf{e}_x$, $\mathbf{u}_2 = \mathbf{e}_y$ and

$$\mathbf{u}_p = -p\,\boldsymbol{\xi}_p$$

With the Rayleigh expansion we obtain

$$\mathbf{A}(0) = -p\,\boldsymbol{\xi}_p(4\pi)^{\frac{1}{2}} \sum_{l=0}^{\infty} i^l(2l+1)^{\frac{1}{2}}\, j_l(kr)\, Y_{l0}(\cos\theta) \qquad (7.36a)$$

---

[7] E. L. Church and J. Weneser, *Phys. Rev.* **100**, 943 (1953); see reference 14, Chapter IV.

where $z = kr \cos \theta$, and to emphasize the special coordinate system the potential has been written as $\mathbf{A}(0)$. Employing the inversion formula (3.8) which is here written in the form

$$Y_{l0}\,\boldsymbol{\xi}_p = \sum_{\lambda} C(l1\lambda; 0p)\,\mathbf{T}_{\lambda lp} \qquad (7.36b)$$

where $\lambda = l, l \pm 1$, we insert the values of the C-coefficients as follows:

$$C(l1l; 0p) = -\frac{p}{\sqrt{2}}, \qquad\qquad l > 0$$

$$C(l1\,l{+}1; 0p) = \left[\frac{l+2}{2(2l+1)}\right]^{\frac{1}{2}}$$

$$C(l1\,l{-}1; 0p) = \left[\frac{l-1}{2(2l+1)}\right]^{\frac{1}{2}}, \qquad l > 0$$

to obtain

$$\mathbf{A}(0) = (2\pi)^{\frac{1}{2}} \sum_{l=0}^{\infty} i^l (2l+1)^{\frac{1}{2}}$$

$$\times j_l \left\{ \mathbf{T}_{llp} - p\left(\frac{l+2}{2l+1}\right)^{\frac{1}{2}}\mathbf{T}_{l+1,lp} - p\left(\frac{l-1}{2l+1}\right)^{\frac{1}{2}}\mathbf{T}_{l-1,lp} \right\} \quad (7.36c)$$

In (7.36c) we replace l by $L - 1$ in the second term, l by $L + 1$ in the third, and change l to L in the first. Then, since in (7.36c) the first term applies for $l > 0$ only and the third vanishes for $l < 2$, all three terms are summed from $L = 1$ to ∞. Consequently, when reference to (7.26) with $C_L(m) = 1$ and to (7.32b) is made, one sees that

$$\mathbf{A}(0) = (2\pi)^{\frac{1}{2}} \sum_{L=1}^{\infty} i^L (2L+1)^{\frac{1}{2}}[\mathbf{A}_{Lp}(m) + ip\,\mathbf{A}_{Lp}(e)] \qquad (7.37)$$

To obtain the potential for an arbitrary direction of propagation with polar angle θ and azimuth angle φ, we make a rotation of $\mathbf{A}(0)$, and, since each term transforms like the components of an irreducible tensor of rank L, it follows that

$$\mathbf{A} = \mathbf{u}_p\, e^{i\mathbf{k}\cdot\mathbf{r}} = (2\pi)^{\frac{1}{2}} \sum_{L=1}^{\infty} \sum_{M=-l}^{L} i^L (2L+1)^{\frac{1}{2}} D_{Mp}^{L}(\varphi\,\theta\,0)$$

$$\times [\mathbf{A}_{LM}(m) + ip\,\mathbf{A}_{LM}(e)] \quad (7.38)$$

This is the expansion referred to in section 19.

So far in this discussion we have implied that the angular momentum associated with a radiation field is L if the vector potential fulfills the eigenvalue equation

$$\mathbf{J}^2 \mathbf{A}_{LM} = L(L + 1) \mathbf{A}_{LM}$$

where $\mathbf{J} = \mathbf{L} + \mathbf{S}$ is the operator of the total angular momentum. Strictly speaking, the angular momentum properties of the electromagnetic field should be discussed in terms of a quantum theory.[8] However, the present (correspondence principle) treatment of the angular momentum properties suffices not only because it gives correct results but also because it is consistent with the conservation of angular momentum in processes involving emission or absorption of photons. That this is so can be seen quite simply.

The transition probability for an emission or absorption process involving a single photon depends on the volume integral of the interaction energy density $\mathbf{j} \cdot \mathbf{A}$ where $\mathbf{j}$ is the current density of the source and, as before, $\mathbf{A}$ is the vector potential of the radiation. For simplicity a term involving the scalar potential is omitted, and this can be justified by using the solenoidal gauge. The quantum-mechanical current density associated with a pair of states ψ_f and ψ_i, describing final and initial states, has the form

$$\mathbf{j} = \psi_f^* \mathbf{v} \psi_i \tag{7.39}$$

where $\mathbf{v}$ is some first-rank tensor. Without loss of generality the vector potential $\mathbf{A}$ may be taken in the form (7.38). It follows then that the transition depends on a sum of terms containing matrix elements of the form

$$(\psi_f | \mathbf{v} \cdot \mathbf{A}_{LM} | \psi_i) \tag{7.40a}$$

where both electric and magnetic multipoles occur.

We now consider that both initial and final states have definite angular momentum, z-projection of angular momentum, and parity. The matrix element (7.40a) becomes

$$(j_f m_f \pi_f | \mathbf{v} \cdot \mathbf{A}_{LM}(\pi_r) | j_i m_i \pi_i) \tag{7.40b}$$

where we have used a notation in which the quantum numbers are placed in evidence. Here π_r refers to the parity of $\mathbf{A}_{LM}$. Referring to the discussion given at the end of Chapter V, it follows that $\mathbf{v} \cdot \mathbf{A}_{LM}$ is an irreducible tensor of rank L with projection quantum number M. Its parity is $(-)^{L+1}\pi_v$ and $(-)^L \pi_v$ for electric and magnetic multipoles, respectively. Here π_v is the parity of the vector $\mathbf{v}$. Since the current density

---

[8] For instance, J. M. Jauch and F. Rohrlich, *Theory of Photons and Electrons*, Addison-Wesley, Cambridge, 1955.

is a polar vector, this gives $\pi_v = -1$. Applying the Wigner–Eckart theorem, it is now clear that only those matrix elements are non-vanishing for which $m_f = M + m_i$, which is the projection quantum number conservation for absorption, and where the following selection rules are fulfilled as well:

For electric transitions, $\quad \Delta(j_f L j_i), \, \pi_f \pi_i = (-)^L$

For magnetic transitions, $\quad \Delta(j_f L j_i), \, \pi_f \pi_i = (-)^{L+1}$

These selection rules express the conservation of angular momentum for absorption (or emission) of radiation of total angular momentum L. They also give the parity selection rule.

Although we have assumed the use of the solenoidal gauge in the foregoing, no results are changed if any other gauge is used. In particular both parity and total angular momentum are not changed by the gauge transformation generated as in (7.31). It is also of interest to note that the intrinsic spin of the photon is a gauge-invariant quantity but that, in general, the orbital angular momentum is not. Of course, for the magnetic multipole only the solenoidal gauge makes sense. For a given total angular momentum L, a gauge transformation introduces terms in the vector potential proportional to $\mathbf{T}_{L,L\pm1,M}$, and these have the wrong parity. Although $\mathbf{A}_{LM}(m)$ is an eigenfunction of $\mathbf{L}^2$, this does not have any direct consequence as far as angular momentum conservation is concerned. In the electric multipoles we have in general a mixture of $\mathbf{T}_{L,L\pm1,M}$, and only for special gauges will one of these disappear; see (7.31a, b).

VIII. STATIC INTERACTIONS

28. MULTIPOLE MOMENTS OF AN ELECTROSTATIC CHARGE DISTRIBUTION

We begin with a rather simple problem of electrostatics which arises in the discussion of the interaction of two systems containing charged particles, for example, two atoms or molecules. It is then possible to write the energy as a sum of terms: dipole–dipole, dipole–quadrupole interactions, etc.[1] These correspond to the fact that each charge distribution will be characterized by a set of multipole moments beginning with the monopole moment (which is just the total charge). In this first example we attempt to find a general expression for the multipole moments.

First consider a charge distribution characterized by L vectors $\mathbf{a}_1$, $\mathbf{a}_2, \cdots, \mathbf{a}_L$ which completely define the positions of the charges. These

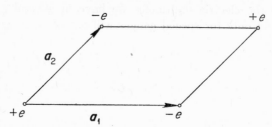

Fig. 8. The elementary quadrupole defined by the vectors $\mathbf{a}_1$ and $\mathbf{a}_2$.

vectors are defined in such a way that the lowest moment that is not zero is the 2^L pole moment. The results given below then serve to provide a more precise definition. Of course, for $L = 1$ we have a dipole moment, and $\mathbf{a}_1$ is the vector separation of the charges. For $L = 2$ we must have a quadrupole moment arising from four charges arranged as shown in Fig. 8. Here the two dipoles, characterized by $\pm \mathbf{a}_1$ and separated by $\mathbf{a}_2$ or by $\pm \mathbf{a}_2$ and separated by $\mathbf{a}_1$, give canceling contributions to the dipole part of the electrostatic potential. In an actual system

[1] H. Margenau, *Revs. Mod. Phys.* **11**, 1 (1939).

the charges are not held rigidly in the specified positions, but appropriate averages over all positions must be taken. However, presumably artificial arrangements such as shown in Fig. 8 are of interest because of the possibility of expressing the electrostatic potential as a sum of terms of the type arising from these arrangements.

Since the potential of a dipole is

$$V_1 = -e\, \mathbf{a}_1 \cdot \nabla \frac{1}{r}$$

and a quadrupole can be formed by putting two dipoles together as illustrated, the corresponding quadrupole potential is

$$V_2 = \frac{e}{2}\, \mathbf{a}_1 \cdot \nabla \, \mathbf{a}_2 \cdot \nabla \frac{1}{r}$$

The extension of this for the 2^L pole charge distribution is

$$V_L = \frac{(-)^L e}{L!} \prod_{n=1}^{L} (\mathbf{a}_n \cdot \nabla) \frac{1}{r} \tag{8.1}$$

It is clear that V_L is symmetric in all the argument vectors $\mathbf{a}_n$.

To evaluate V_L, we use the gradient formula (6.42). First notice that in calculating

$$\nabla \frac{1}{r} = (4\pi)^{\frac{1}{2}} \nabla \frac{1}{r} Y_{00} = \frac{4\pi}{r^2} \mathbf{T}_{010}$$

only the first term of (6.42) contributes. This will continue to be the case as the operators $\mathbf{a}_n \cdot \nabla$ in (8.1) are successively applied. This is obvious if we recognize that

$$\prod_{1}^{N} \mathbf{a}_n \cdot \nabla \frac{1}{r}$$

where N is arbitrary, is a solution of Laplace's equation, irregular at $r = 0$, and must contain the radial factor r^{-N-1} and, therefore, the factor Y_{Nm}. It is clear that the second term of (6.42) always vanishes when applied to such a term.

We observe that

$$\mathbf{a}_1 \cdot \nabla \frac{1}{r} = \frac{(4\pi)^{\frac{1}{2}}}{r^2} \sum_{\mu} C(110; \mu, -\mu)\, Y_{1,-\mu} \mathbf{a}_1 \cdot \boldsymbol{\xi}_{\mu}$$

$$= -\frac{4\pi}{3r^2} \sum_{\mu} (-)^{\mu}\, Y_{1,-\mu}\, \mathcal{Y}_{1\mu}(\mathbf{a}_1) \tag{8.2}$$

where $\mathcal{Y}_{1\mu}(\mathbf{a}_1) = a_1 Y_{1\mu}(\mathbf{a}_1)$ is the solid harmonic of degree unity in the space of $\mathbf{a}_1$. Equation (8.2) is recognized to be an invariant (zero-rank tensor) as it must be since $\mathbf{a}_1 \cdot \nabla$ is a scalar operator. Similarly V_L must be a zero-rank tensor. Therefore from what has been said it is possible to write it in the form

$$V_L = -r^{-L-1} \sum_M (-)^M B_{LM}(\mathbf{a}_1 \cdots \mathbf{a}_L) Y_{L,-M} \qquad (8.3)$$

where the B_{LM} are irreducible tensors of rank L and

$$V_N = e \prod_1^N \left(-\frac{\mathbf{a}_n}{n} \cdot \nabla\right) \frac{1}{r}$$

$$= -r^{-N-1} \sum_M (-)^M B_{NM}(\mathbf{a}_1 \cdots \mathbf{a}_N) Y_{N,-M} \qquad (8.3a)$$

From (8.2) we know that

$$B_{1M} = -\frac{4\pi e}{3} \mathcal{Y}_{1M}(\mathbf{a}_1) \qquad (8.4)$$

The tensors B_{LM} will constitute the 2^L pole moments. The simplest way to determine them is by a recurrence relation giving $B_{N+1,M}$ in terms of the $2N+1$ components B_{NM}, and then (8.4) determines all of them. Thus,

$$V_{N+1} = -r^{-N-2} \sum_M (-)^M B_{N+1,M}(\mathbf{a}_1 \cdots \mathbf{a}_{N+1}) Y_{N+1,-M}$$

$$= -\frac{\mathbf{a}_{N+1}}{N+1} \cdot \nabla V_N$$

$$= \frac{\mathbf{a}_{N+1}}{N+1} \cdot \sum_M (-)^M B_{NM}(\mathbf{a}_1 \cdots \mathbf{a}_N) \nabla r^{-N-1} Y_{N,-M} \qquad (8.5)$$

From the gradient formula (6.42)

$$\nabla r^{-N-1} Y_{N,-M} = [(N+1)(2N+1)]^{\frac{1}{2}} r^{-N-2} \mathbf{T}_{N,N+1,-M}$$

Therefore

$$V_{N+1} = \left(\frac{2N+1}{N+1}\right)^{\frac{1}{2}} r^{-N-2} \sum_M (-)^M B_{NM}$$

$$\times \sum_\mu C(1\ N+1\ N; -\mu, -M+\mu) Y_{N+1,-M+\mu} \mathbf{a}_{N+1} \cdot \mathbf{\xi}_{-\mu} \qquad (8\ 6)$$

In (8.6) change $M - \mu$ to M' and use

$$\mathbf{a}_{N+1} \cdot \mathbf{\xi}_{-\mu} = \left(\frac{4\pi}{3}\right)^{\frac{1}{2}} \mathcal{Y}_{1,-\mu}(\mathbf{a}_{N+1})$$

Then

$$V_{N+1} = \left[\frac{4\pi}{3}\frac{2N+1}{N+1}\right]^{\frac{1}{2}} r^{-N-2} \sum_{M'} (-)^{M'} Y_{N+1,-M'}$$
$$\times \sum_{\mu} (-)^{\mu} C(1\ N+1\ N; \mu M') B_{N,M'+\mu} \mathcal{Y}_{1,-\mu}(\mathbf{a}_{N+1}) \quad (8.6a)$$

Comparison of (8.6a) and (8.5) gives the recurrence formula

$$B_{N+1,M}(\mathbf{a}_1 \cdots \mathbf{a}_{N+1}) = -\left[\frac{4\pi}{3}\frac{2N+1}{N+1}\right]^{\frac{1}{2}}$$
$$\times \sum_{\mu} (-)^{\mu} C(1\ N+1\ N; \mu M) B_{N,M+\mu}(\mathbf{a}_1 \cdots \mathbf{a}_N) \mathcal{Y}_{1,-\mu}(\mathbf{a}_{N+1}) \quad (8.7)$$

Since the sum in (8.7) can be written in the form

$$-\left(\frac{2N+1}{2N+3}\right)^{\frac{1}{2}} \sum_{\mu} C(1N\ N+1; -\mu, \mu+M)$$
$$\times B_{N,\mu+M}(\mathbf{a}_1 \cdots \mathbf{a}_N) \mathcal{Y}_{1,-\mu}(\mathbf{a}_{N+1})$$

it follows that $B_{N+1,M}$ are the components of a tensor of rank $N+1$, as was to be expected.

If we deal with a continuous charge distribution, then we set $\mathbf{a}_1 = \mathbf{a}_2 = \cdots \mathbf{a}_L = \mathbf{r}'$ and eventually integrate over $\mathbf{r}'$ after replacing e by the element of charge $\rho\ d\mathbf{r}'$. In this case we should get the standard result. As a check we investigate this point. We set

$$B_{NM} = b_N Y_{NM}(\mathbf{r}') \quad (8.8)$$

where b_N is independent of M. Then substitution in (8.7) gives

$$b_{N+1} Y_{N+1,M} = -\left[\frac{4\pi}{3}\frac{2N+1}{N+1}\right]^{\frac{1}{2}} b_N r'$$
$$\times \sum_{\mu} (-)^{\mu} C(1\ N+1\ N; \mu M) Y_{N,M+\mu} Y_{1,-\mu}$$

and the argument of all spherical harmonics is $\mathbf{r}'$. Using the coupling rule (4.32), this becomes

$$b_{N+1} Y_{N+1,M} = -\left[\frac{4\pi}{3}\frac{2N+1}{N+1}\right]^{\frac{1}{2}} r' b_N \sum_{\lambda\mu} (-)^{\mu} \left[\frac{3(2N+1)}{4\pi(2\lambda+1)}\right]^{\frac{1}{2}}$$
$$\times C(1\ N+1\ N; \mu M) C(1N\lambda; -\mu, \mu+M) C(1N\lambda; 00) Y_{\lambda M} \quad (8.9)$$

The sum over μ is

$$\left(\frac{2N+1}{2N+3}\right)^{\frac{1}{2}} \sum_{\mu} (-)^{\mu}(-)^{1-\mu} C(1N\ N+1; -\mu, \mu+M)$$
$$\times C(1N\lambda; -\mu, \mu+M) = -\left(\frac{2N+1}{2N+3}\right)^{\frac{1}{2}} \delta_{\lambda,N+1}$$

after the symmetry rules (3.16c) and (3.16a) have been used. The fact that only $Y_{N+1,M}$ occurs on the right side of (8.9) justifies the form (8.8). Now we get

$$b_{N+1} = \left(\frac{2N+1}{N+1}\right)^{\frac{1}{2}} \frac{2N+1}{2N+3} C(1N\ N+1; 00)\ r'b_N$$

$$= \frac{2N+1}{2N+3} r'b_N \tag{8.10}$$

From (8.4) we see that

$$b_1 = -\frac{4\pi}{3} er'$$

so that (8.10) gives

$$b_L = -\frac{4\pi er'^L}{2L+1} \tag{8.11}$$

and

$$V_L = \frac{4\pi er'^L}{(2L+1)r^{L+1}} \sum_M (-)^M\ Y_{1M}(\mathbf{r'})\ Y_{1,-M}(\mathbf{r})$$

$$= \frac{er'^L}{r^{L+1}} P_L(\cos\Theta) \tag{8.12}$$

In (8.12) Θ is the angle between the vectors $\mathbf{r'}$ and $\mathbf{r}$. This result is just the standard one obtained by expanding $e/|\mathbf{r} - \mathbf{r'}|$ for $r > r'$ and taking the coefficient of P_L.

29. SPIN INTERACTIONS

There is an important connection between the electromagnetic moments of a quantum-mechanical system and its spin. The interaction between such systems is essentially a coupling of angular momenta. We see this in a formal way in the following manner. Consider the interaction between two systems with total angular momenta I and J. Assume that the systems have Lth-order multipole moments. Then the interaction energy due to these moments is a contraction of two tensors of rank L. Each one of these tensors must be expressed in the space of its respective angular momentum vector. Now there is essentially only one Lth-rank tensor that can be constructed out of the components of the vector operator $\mathbf{I}$. This is

$$\mathfrak{I}_{LM}(\mathbf{I}) = (\mathbf{I} \cdot \nabla)^L\ \mathcal{Y}_{LM}(\mathbf{r}) \tag{8.13}$$

Since $\mathbf{I} \cdot \nabla$ is a scalar operator the same is true of $(\mathbf{I} \cdot \nabla)^L$ and, hence, $\mathfrak{I}_{LM}$, just as $\mathcal{Y}_{LM}$, is a tensor of rank L. The process of differentiation and scalar multiplication by $\mathbf{I}$ is one that eventually replaces each component of $\mathbf{r}$ by the corresponding component of $\mathbf{I}$. Thus, where $\mathcal{Y}_{LM}$ is a homogeneous function of degree L in the components of $\mathbf{r}$, the tensor $\mathfrak{I}_{LM}(\mathbf{I})$ is homogeneous of degree L in the components of $\mathbf{I}$.[2] There is one distinction: the components of $\mathbf{r}$ commute, and those of $\mathbf{I}$ do not. Instead

$$I_\mu I_\nu - I_\nu I_\mu = \pm I_{\mu+\nu} \qquad (\mu, \nu = \pm 1, 0) \qquad (8.14)$$

In (8.14) the upper sign goes with $\mu < \nu$ and the lower with $\mu > \nu$.

We can obtain a clue as to the structure of the tensors $\mathfrak{I}_{LM}$ if we note that

$$\nabla \mathcal{Y}_{LM} = \nabla r^L\, Y_{LM} = [L(2L+1)]^{\frac{1}{2}} r^{L-1}\, \mathbf{T}_{L,L-1,M}$$

$$= [L(2L+1)]^{\frac{1}{2}} r^{L-1} \sum_\mu C(1\ L-1\ L;\ -\mu,\, \mu+M)\, Y_{L-1,M+\mu}\, \boldsymbol{\xi}_{-\mu}$$

$$= [L(2L+1)]^{\frac{1}{2}} \sum_\mu C(1\ L-1\ L;\ -\mu,\, \mu+M)\, \mathcal{Y}_{L-1,M+\mu}\, \boldsymbol{\xi}_{-\mu}$$

$$(8.15)$$

Thus the gradient operation on the regular solid harmonics generates a sum of solid harmonics of degree one lower. This suggests that we define a sequence of tensors $t_{\lambda M}$ by the recurrence relation

$$t_{\lambda M}(\mathbf{I}) = \sum_\mu C(1\ \lambda-1\ \lambda;\ -\mu,\, \mu+M)\, t_{\lambda-1,\mu+M}(\mathbf{I})\, \mathcal{Y}_{1,-\mu}(\mathbf{I}) \qquad (8.16)$$

and

$$t_{1M} = \mathcal{Y}_{1M}; \qquad t_{00} = 1$$

If we now set

$$\mathfrak{I}_{LM}(\mathbf{I}) = A_L t_{LM} \qquad (8.17)$$

where A_L is a constant, it is easy to see by application of (8.15) that

$$A_L = \left(\frac{4\pi}{3}\right)^{L/2} \left[\frac{L!(2L+1)!!}{4\pi}\right]^{\frac{1}{2}} \qquad (8.17a)$$

and

$$(2L+1)!! = 1\cdot 3\cdot 5\, \cdots\, (2L+1)$$

As an example

$$\mathfrak{I}_{2,\pm2}(\mathbf{I}) = \left(\frac{15}{2\pi}\right)^{\frac{1}{2}} I^2_{\pm1}$$

[2] This process is known as "polarization," and the resulting tensors are sometimes referred to as polarized harmonics; see H. Weyl, *The Classical Groups*, Princeton University Press, 1939. Also D. L. Falkoff and G. E. Uhlenbeck, *Phys. Rev.* **79**, 323 (1950).

$$\Im_{2,\pm 1}(\mathbf{I}) = \left(\frac{15}{4\pi}\right)^{\frac{1}{2}} (I_0 I_{\pm 1} + I_{\pm 1} I_0)$$

$$\Im_{20}(\mathbf{I}) = \left(\frac{5}{4\pi}\right)^{\frac{1}{2}} (3I_0^2 - \mathbf{I}^2) \tag{8.18}$$

Notice the symmetrization, for example, in $\Im_{2,\pm 1}$.

For the two systems referred to above, the interaction energy is

$$\mathfrak{IC} = K \sum_M (-)^M \Im_{LM}(\mathbf{I}) \Im_{L,-M}(\mathbf{J}) \tag{8.19}$$

where K is some constant. This form of the energy can be expressed alternatively in terms of $\mathbf{I}$ and $\mathbf{J}$ for each value of L. For instance, with $L = 1$ we get

$$\mathfrak{IC} = \frac{3K}{4\pi} \mathbf{I} \cdot \mathbf{J}$$

which is the familiar dipole–dipole coupling leading to fine structure or hyperfine structure.

For $L = 2$, direct substitution of (8.18) can be made, but it is much easier to carry out the calculation by evaluating the matrix elements of $\mathfrak{IC}$. For this purpose we evaluate

$$(Im_i \,|\, \Im_{LM} \,|\, I'm_i') = C(I'LI; m_i'M) \,(I \,\|\, \Im_L \,\|\, I')$$

by the Wigner–Eckart theorem. We can evaluate this by considering a special case: $M = L$, $m_i = I$, and therefore $m_i' = I - L$. In (8.16) with $\lambda = M = L$, only $\mu = -1$ can contribute since $\mu + M \leq L - 1$. Thus,

$$t_{LL} = C(1\ L-1\ L; 1, L-1)\ t_{L-1,L-1}(\mathbf{I})\ \mathcal{Y}_{11}(\mathbf{I})$$

$$= t_{L-1,L-1}(\mathbf{I})\ \mathcal{Y}_{11}(\mathbf{I}) = [\mathcal{Y}_{11}(\mathbf{I})]^L$$

But

$$(Im \,|\, \mathcal{Y}_{11}(\mathbf{I}) \,|\, I'm') = -[(3/8\pi)(I - m')(I + m' + 1)]^{\frac{1}{2}}\, \delta_{II'}\, \delta_{m,m'+1}$$

and using this

$$(II \,|\, [\mathcal{Y}_{11}(\mathbf{I})]^L \,|\, I\ I-L)$$

$$= (II \,|\, \mathcal{Y}_{11} \,|\, I\ I-1)(I\ I-1 \,|\, \mathcal{Y}_{11} \,|\, I\ I-2) \cdots (I\ I-L+1 \,|\, \mathcal{Y}_{11} \,|\, I\ I-L)$$

$$= (-)^L \left(\frac{3}{8\pi}\right)^{L/2} \left[\frac{L!(2I)!}{(2I - L)!}\right]^{\frac{1}{2}} \tag{8.20}$$

It is now necessary to evaluate $C(ILI; I-L, L)$. From (3.18) we see that the summation index $\varkappa$ is confined to a single value: $\varkappa = L$ since the denominator of (3.18) contains $(\varkappa - L)!(L - \varkappa)!$. Consequently,

$$C(ILI; I-L, L) = (-)^L \left[\frac{(2I + 1)!(2L)!}{(2I + L + 1)!L!} \right]^{\frac{1}{2}} \qquad (8.21)$$

Combining (8.20) and (8.21), we find

$$(Im_i | \mathcal{I}_{LM} | I'm_i') = \delta_{m_i, m_i' + M} \, \delta_{II'} \frac{L!}{2^L}$$

$$\times \left[\frac{1}{4\pi} \frac{2L + 1}{2I + 1} \frac{(2I + L + 1)!}{(2I - L)!} \right]^{\frac{1}{2}} C(ILI; m_i'M) \quad (8.22)$$

Incidentally, this result shows that for given spin I only those multipole moments exist for which $L \leq 2I$.

We are now in a position to consider the matrix elements of $\mathcal{K}$. We consider a representation in which $\mathbf{F}^2$ and F_z are diagonal, where $\mathbf{F} = \mathbf{I} + \mathbf{J}$ is the total angular momentum. Clearly $\mathcal{K}$ is diagonal in F and μ, the eigenvalue of F_z. Then from (6.22)

$$(F\mu | \mathcal{K} | F\mu) = K(-)^{F-I-J} [(2I + 1)(2J + 1)]^{\frac{1}{2}} W(JJII; LF)$$
$$\times (I \parallel \mathcal{I}_L(I) \parallel I)(J \parallel \mathcal{I}_L(J) \parallel J)$$

We specialize to the case $L = 2$ which will yield the quadrupole energy. Then

$$(I \parallel \mathcal{I}_2(I) \parallel I) = \frac{1}{2} \left[\frac{5}{4\pi} \frac{(2I + 3)!}{(2I - 1)!} \right]^{\frac{1}{2}}$$

and similarly for $[J \parallel \mathcal{I}_2(J) \parallel J]$. From reference 3,

$$W(JJII; 2F) = 6(-)^{F-I-J} \left[\frac{(2J + 3)!}{(2J - 1)!} \frac{(2I + 3)!}{(2I - 1)!} \right]^{-\frac{1}{2}}$$
$$\times [A(A + 1) - \tfrac{4}{3}I(I + 1)J(J + 1)]$$

where $A = F(F + 1) - I(I + 1) - J(J + 1)$.

Obviously A is the diagonal matrix element of $2\mathbf{I} \cdot \mathbf{J}$. Thus we find

$$(F\mu | \mathcal{K} | F\mu) = (F\mu | \mathcal{E} | F\mu)$$

where the operator $\mathcal{E}$ is

$$\mathcal{E} = \frac{15K}{2\pi} \left[(\mathbf{I} \cdot \mathbf{J})^2 + \frac{1}{2} \mathbf{I} \cdot \mathbf{J} - \frac{1}{3} \mathbf{I}^2 \mathbf{J}^2 \right] \qquad (8.23)$$

[3] L. C. Biedenharn, *Oak Ridge Natl. Lab. Rept. 1098.*

This is the form of the energy operator as it often appears in the literature.[4] For higher multipoles similar results can be obtained directly from (8.22) and Appendix I, B2.

For some purposes, for example, the study of the energy levels of a molecule in a strong magnetic field when there is also magnetic dipole or electric quadrupole coupling, we find the matrix elements in the uncoupled representation useful.[4] These are obtained immediately from (8.22):

$$(m_i m_j | \mathfrak{K} | m_i' m_j') = K(-)^{m_i - m_i'} \delta_{m_i - m_i', m_j' - m_j}$$

$$\times \left(\frac{L!}{2^L}\right)^2 \frac{(2L+1)}{[(2I+1)(2J+1)]^{\frac{1}{2}}} \left[\frac{(2I+L+1)}{(2I-L)!} \frac{(2J+L+1)!}{(2J-L)!}\right]^{\frac{1}{2}}$$

$$\times C(ILI; m_i', m_i - m_i') \, C(JLJ; m_j, m_j' - m_j)$$

and

$$-L \leq m_i - m_i' \leq L \qquad (8.24)$$

In most applications the coupling energy (8.19) is a small term to be added to the rest of the Hamiltonian H_0. In that case the interaction term $\mathfrak{K}$ can be treated as a small perturbation. The unperturbed solutions are eigenfunctions of H_0 and are identical with the decoupled representation. For the sake of definiteness we refer to the angular momentum J as arising from electrons which are coupled to the nucleus with angular momentum (spin) I. Quite often we are not interested in the electronic state, and no transitions between different unperturbed electronic states occur. Under these circumstances it is appropriate to average over this unperturbed electronic state. This is equivalent to using first-order perturbation in the J-space. The diagonal matrix elements of the interaction $\mathfrak{K}$ is then

$$(m_j | \mathfrak{K} | m_j) = K(m_j | \mathfrak{J}_{L0}(\mathbf{J}) | m_j) \mathfrak{J}_{L0}(\mathbf{I}) \qquad (8.25)$$

since $m_j + M = m_j$, or $M = 0$.

[4] For example, B. T. Feld and W. E. Lamb Jr., Phys. Rev. 67, 15 (1945). Comparison with their equation II(11) gives the significance of the constant K. One finds

$$K = \frac{\pi}{5} \frac{e^2 qQ}{I(2I-1) J(2J-1)}$$

where $e^2 qQ$ is the quantity generally referred to as the quadrupole coupling energy: Q is the nuclear quadrupole moment (in square centimeters) and q [which has the dimensions of (length)$^{-3}$] is essentially the gradient of the electric field at the position of the nucleus.

For $L = 2$ this yields the very familiar result

$$(m_j | \mathfrak{K} | m_j) = \frac{5K}{4\pi} (m_j | 3J_0^2 - \mathbf{J}^2 | m_j)(3I_0^2 - \mathbf{I}^2)$$

It will be recognized that the tensors $\mathfrak{I}_{L0}$ bear a certain resemblance to the Legendre polynomials, or better, to the solid harmonics $\mathcal{Y}_{L0} \sim r^L P_L$. Thus, in (8.18) we obtain $\mathfrak{I}_{20}$, within a multiplicative factor, by replacing the coordinates z and r^2 by I_0 and $\mathbf{I}^2$, respectively. However, while this procedure is correct for $L = 2$, it is not generally applicable. Thus, from (8.16), (8.17a), and (8.18) we find for the octupole operator

$$\mathfrak{I}_{30}(\mathbf{I}) = 3 \left(\frac{7}{4\pi} \right)^{\frac{1}{2}} I_0 [5I_0^2 - 3\mathbf{I}^2 + 1] \qquad (8.26)$$

$\mathfrak{I}_{30}$ would be proportional to $\mathcal{Y}_{30}$, with the replacements described above, only if the last term in (8.26) is ignored. This last term arises directly from the commutator of I_1 and I_{-1} and is of quantum-mechanical origin. The correspondence of these tensors and the spherical harmonics applies only in the classical limit where the angular momenta are large.

IX. PARTICLES OF SPIN $\frac{1}{2}$

30. NON-RELATIVISTIC DESCRIPTION

Since a large part of the interest in problems dealing with angular momentum arises in nuclear physics, it is important to study the properties of particles of spin $\frac{1}{2}$, for instance, the neutrons and protons forming the nucleus. As far as nuclear physics is concerned, the results of this section can be applied to the simplified j–j coupling model with one particle outside of a closed shell. Such applications will be considered in Chapter XI.

We are concerned with the description of a particle in a central field. As is well known,[1] the spin gives rise to a spin-orbit coupling of the form

$$\phi(r)\,\mathbf{S} \cdot \mathbf{L} \tag{9.1}$$

where for charged particles in an electrostatic field

$$\phi(r) = \frac{\hbar^2}{2m^2c^2} \frac{1}{r} \frac{dV}{dr} \tag{9.2}$$

in which V is the central potential. The nature of the spin-orbit coupling in nuclei is not known. Nevertheless, if we adopt (9.1) as a model expressing the spin dependence of the coupling, the fact that $\phi(r)$ is not specified simply means that a scale parameter is involved in the theory. Since neither S_z or L_z commutes with the spin-orbit energy but $J_z = S_z + L_z$ does, it follows that J_z is diagonal but the z-components of intrinsic and orbital angular momentum are not separately diagonal. Also $\mathbf{J}^2$, $\mathbf{L}^2$, and $\mathbf{S}^2$ commute with (9.1) and are simultaneously diagonal with the energy H. Accordingly, a wave function which diagonalizes these operators is

$$\Psi = R(r)\,\chi_{\varkappa\mu} \tag{9.3}$$

where

$$\chi_{\varkappa\mu} = \sum_m C(l\tfrac{1}{2}j;\, m,\, \mu - m)\, Y_{lm}\, \psi_{\frac{1}{2},\mu - m}; \qquad \mu - m = \pm\tfrac{1}{2} \tag{9.4}$$

[1] For example, L. Schiff, *Quantum Mechanics*, 2nd edition, p. 286, McGraw-Hill Book Co., New York, 1955.

150

with the intrinsic spin functions $\psi_{\frac{1}{2}m_s}$ defined by (4.47), where only the m_s index was written. We have introduced the symbol $\varkappa$ which gives both j and l according to

$$j = |\varkappa| - \tfrac{1}{2}$$
$$l = \varkappa \qquad \text{if} \quad \varkappa > 0 \qquad\qquad (9.5)$$
$$= -\varkappa - 1 \quad \text{if} \quad \varkappa < 0$$

$\varkappa = -1, 1, -2, 2, \cdots$ for $s_{\frac{1}{2}}, p_{\frac{1}{2}}, p_{\frac{3}{2}}, d_{\frac{3}{2}}, \cdots$ particles, respectively. In general, $\varkappa \neq 0$. We could write $l \equiv l_\varkappa$ (since it depends on $\varkappa$) in the form

$$l = j + \tfrac{1}{2}S(\varkappa)$$

where $S(\varkappa) = \pm 1$ is the sign of $\varkappa$. Also

$$l_\varkappa = l_{-\varkappa} + S(\varkappa)$$

Clearly

$$\mathbf{S} \cdot \mathbf{L} \chi_{\varkappa\mu} = \tfrac{1}{2}[\mathbf{J}^2 - \mathbf{L}^2 - \mathbf{S}^2]\chi_{\varkappa\mu} = \tfrac{1}{2}[j(j+1) - l(l+1) - \tfrac{3}{4}]\chi_{\varkappa\mu}$$
$$= -\tfrac{1}{2}(\varkappa + 1)\chi_{\varkappa\mu}$$

Substitution of (9.3) into the Schrödinger equation

$$H\Psi = E\Psi = \left[-\frac{\hbar^2}{2m}\nabla^2 + V + \phi(r)\, \mathbf{S}\cdot\mathbf{L} \right]\Psi$$

together with

$$\nabla^2 = \nabla_r^2 - \frac{1}{r^2}L^2; \qquad \nabla_r^2 = \frac{\partial^2}{\partial r^2} + \frac{2}{r}\frac{\partial}{\partial r}$$

gives the wave equation

$$\nabla_r^2 R + \frac{2m}{\hbar^2}\left\{ E - V - \frac{\hbar^2}{2m}\left[\frac{l(l+1)}{r^2}\right] + \frac{1}{2}\phi(r)(1+\varkappa) \right\} R = 0 \quad (9.6)$$

for the radial wave function R.

The quantum numbers for the diagonal operators are $j, l, s = \frac{1}{2}$, and μ where the latter is the eigenvalue of j_z. The two component spin-angular function $\chi_{\varkappa\mu}$, which transforms with D^j, is a spinor wave function; see section 15. In particular, for an $s_{\frac{1}{2}}$ particle

$$\chi_{-1\,\mu} = \psi_{\frac{1}{2}\mu} Y_{00}$$

and for a $p_{\frac{1}{2}}$ particle

$$\chi_{1\mu} = \begin{bmatrix} -(\tfrac{1}{2} - \tfrac{1}{3}\mu)^{\frac{1}{2}} & Y_{1,\mu-\frac{1}{2}} \\ (\tfrac{1}{2} + \tfrac{1}{3}\mu)^{\frac{1}{2}} & Y_{1,\mu+\frac{1}{2}} \end{bmatrix}$$

Generally,

$$\chi_{\kappa\mu} = C(l\tfrac{1}{2}j; \mu-\tfrac{1}{2}, \tfrac{1}{2})\, Y_{l,\mu-\frac{1}{2}} \begin{pmatrix} 1 \\ 0 \end{pmatrix} + C(l\tfrac{1}{2}j; \mu+\tfrac{1}{2}, -\tfrac{1}{2})\, Y_{l,\mu+\frac{1}{2}} \begin{pmatrix} 0 \\ 1 \end{pmatrix}$$

The wave functions (9.3) are just the single-particle functions in the j–j coupling model of the nucleus. The wave functions for complicated configurations are constructed, using these single-particle wave functions, with the additional requirement that the total wave function diagonalizes the total angular momentum and z-component. For identical fermions the additional requirement of overall antisymmetry is imposed. Two-particle configurations were referred to in section 24. Three-particle configurations of identical particles are considered in Chapter XII.

31. RELATIVISTIC DESCRIPTION

We turn next to the description of a relativistic electron in a central field. As is discussed in any standard treatise on quantum mechanics,[1,2] the equation of motion of such an electron is

$$H\Psi = W\Psi = [-(\boldsymbol{\alpha}\cdot\mathbf{p} + \beta) + V(r)]\Psi \qquad (9.7)$$

for a stationary state of total energy W. Here $\mathbf{p} = -i\nabla$, the potential energy is $V(r)$ and the 4-by-4 matrices $\boldsymbol{\alpha}$ and β can be represented by

$$\boldsymbol{\alpha} = \begin{pmatrix} 0 & \sigma \\ \sigma & 0 \end{pmatrix} \qquad \beta = \begin{pmatrix} I & 0 \\ 0 & -I \end{pmatrix} \qquad (9.8)$$

In (9.8) each element is a 2-by-2 matrix. For example, in β, the I is a unit matrix and 0 a null matrix. Written out in full detail, β and α_x would be

$$\beta = \begin{bmatrix} 1 & 0 & 0 & 0 \\ 0 & 1 & 0 & 0 \\ 0 & 0 & -1 & 0 \\ 0 & 0 & 0 & -1 \end{bmatrix} \qquad \alpha_x = \begin{bmatrix} 0 & 0 & 0 & 1 \\ 0 & 0 & 1 & 0 \\ 0 & 1 & 0 & 0 \\ 1 & 0 & 0 & 0 \end{bmatrix}$$

The units are such that $\hbar = m = c = 1$.

We wish to construct a central field wave function which corresponds to a diagonalization of $\mathbf{J}^2$ and J_z where $\mathbf{J}$ as usual is the total angular momentum operator. For the purpose in hand we must separate the equation into polar coordinates which is done as follows: First, with

$$\nabla = \mathbf{r}(\mathbf{r}\cdot\nabla) - \mathbf{r}\times(\mathbf{r}\times\nabla)$$

$$= \mathbf{r}\frac{\partial}{\partial r} - i\frac{\mathbf{r}}{r}\times\mathbf{L}$$

[2] P. A. M. Dirac, *Quantum Mechanics*, 2nd edition, p. 46, Clarendon Press, Oxford, 1935.

we have

$$\boldsymbol{\alpha} \cdot \nabla = \alpha_r \frac{\partial}{\partial r} - \frac{i}{r} \boldsymbol{\alpha} \cdot \mathfrak{r} \times \mathbf{L} \qquad (9.9)$$

Next we use the result that

$$\boldsymbol{\sigma} \cdot \mathbf{A} \ \boldsymbol{\sigma} \cdot \mathbf{B} = \mathbf{A} \cdot \mathbf{B} + i \boldsymbol{\sigma} \cdot (\mathbf{A} \times \mathbf{B})$$

for any vectors $\mathbf{A}$ and $\mathbf{B}$ whose components commute with the components of $\boldsymbol{\sigma}$. This follows from the commutation rules for the components of $\boldsymbol{\sigma}$:

$$\sigma_j \sigma_k = i \sigma_l$$

where j, k, l is a cyclic permutation of x, y, z. Therefore

$$\boldsymbol{\alpha} \cdot \mathbf{A} \ \boldsymbol{\alpha} \cdot \mathbf{B} = \begin{pmatrix} \mathbf{A} \cdot \mathbf{B} & 0 \\ 0 & \mathbf{A} \cdot \mathbf{B} \end{pmatrix} + i \begin{pmatrix} \boldsymbol{\sigma} \cdot \mathbf{A} \times \mathbf{B} & 0 \\ 0 & \boldsymbol{\sigma} \cdot \mathbf{A} \times \mathbf{B} \end{pmatrix}$$

Let $\mathbf{A} = \mathfrak{r}$ and $\mathbf{B} = \mathbf{L}$. Then $\mathbf{A} \cdot \mathbf{B} = 0$ and

$$\boldsymbol{\alpha} \cdot \mathfrak{r} \ \boldsymbol{\alpha} \cdot \mathbf{L} = i \rho_3 \begin{pmatrix} 0 & \boldsymbol{\sigma} \cdot \mathfrak{r} \times \mathbf{L} \\ \boldsymbol{\sigma} \cdot \mathfrak{r} \times \mathbf{L} & 0 \end{pmatrix} = i \rho_3 \boldsymbol{\alpha} \cdot \mathfrak{r} \times \mathbf{L}$$

where ρ_3 is the 4-by-4 matrix

$$\rho_3 = \begin{pmatrix} 0 & I \\ I & 0 \end{pmatrix}$$

Since ρ_3 and $\boldsymbol{\alpha}$ commute and $\rho_3^{-1} = \rho_3$, we find

$$i \, \boldsymbol{\alpha} \cdot \mathfrak{r} \times \mathbf{L} = \rho_3 \, \boldsymbol{\alpha} \cdot \mathfrak{r} \, \boldsymbol{\alpha} \cdot \mathbf{L} = \boldsymbol{\alpha} \cdot \mathfrak{r} \, \boldsymbol{\sigma} \cdot \mathbf{L} = \alpha_r \begin{pmatrix} \boldsymbol{\sigma} \cdot \mathbf{L} & 0 \\ 0 & \boldsymbol{\sigma} \cdot \mathbf{L} \end{pmatrix}$$

where we have followed conventional usage in writing [3]

$$\rho_3 \boldsymbol{\alpha} = \boldsymbol{\sigma} = \begin{pmatrix} \boldsymbol{\sigma} & 0 \\ 0 & \boldsymbol{\sigma} \end{pmatrix}$$

The context will distinguish between the 2-by-2 Pauli spin matrices and the 4-by-4 $\boldsymbol{\sigma}$-matrices which are the Pauli spin matrices iterated.

The result is

$$\boldsymbol{\alpha} \cdot \nabla = \alpha_r \left(\frac{\partial}{\partial r} - \frac{1}{r} \boldsymbol{\sigma} \cdot \mathbf{L} \right) \qquad (9.10)$$

---

[3] Note

$$\boldsymbol{\alpha} \times \boldsymbol{\alpha} = \begin{pmatrix} \boldsymbol{\sigma} \times \boldsymbol{\sigma} & 0 \\ 0 & \boldsymbol{\sigma} \times \boldsymbol{\sigma} \end{pmatrix} = 2i \begin{pmatrix} \boldsymbol{\sigma} & 0 \\ 0 & \boldsymbol{\sigma} \end{pmatrix} = 2i\boldsymbol{\sigma}$$

We now introduce the wave function in the form

$$\Psi = \begin{pmatrix} -i f \chi_{\varkappa,-\mu} \\ g \chi_{\varkappa\mu} \end{pmatrix} \tag{9.11}$$

where f and g are radial wave functions. The minus sign is a matter of definition, and the i is inserted as a factor of f so that f and g can both be real. This will be seen in the sequel. The $\chi_{\pm\varkappa,\mu}$ are the two component Pauli spinors introduced in section 30.

When (9.11) is introduced in (9.7) and (9.10) is used, we obtain

$$(W - V) \begin{pmatrix} -i f \chi_{-\varkappa,\mu} \\ g \chi_{\varkappa\mu} \end{pmatrix} = i \begin{pmatrix} g' \sigma_r \chi_{\varkappa\mu} \\ -i f' \sigma_r \chi_{-\varkappa,\mu} \end{pmatrix}$$

$$- \frac{i}{r} \begin{pmatrix} g \sigma_r \boldsymbol{\sigma} \cdot \mathbf{L} \chi_{\varkappa\mu} \\ -i f \sigma_r \boldsymbol{\sigma} \cdot \mathbf{L} \chi_{-\varkappa,\mu} \end{pmatrix} - \begin{pmatrix} -i f \chi_{-\varkappa,\mu} \\ -g \chi_{\varkappa\mu} \end{pmatrix} \tag{9.12}$$

Here the prime means differentiation with respect to r. Now, according to the foregoing

$$(1 + \boldsymbol{\sigma} \cdot \mathbf{L}) \chi_{\varkappa\mu} = -\varkappa \chi_{\varkappa\mu} \tag{9.12a}$$

In addition we must also evaluate $\sigma_r \chi_{\varkappa\mu}$. To accomplish this we write

$$\sigma_r \chi_{\varkappa\mu} = \sum_{\varkappa'\mu'} (\chi_{\varkappa'\mu'} | \sigma_r | \chi_{\varkappa\mu}) \chi_{\varkappa'\mu'}$$

Since σ_r is a scalar product of two first-rank tensors, $\mathbf{r}$ and $\boldsymbol{\sigma}$, we can use (6.21) to determine the matrix elements of σ_r. This gives $\mu' = \mu$ and $j' = j$, or $|\varkappa'| = |\varkappa|$. Since σ_r has odd parity, it follows that $l' \neq l$ so that $\varkappa' = -\varkappa$. Thus $l' = l_{-\varkappa}$. The single-matrix element $(\chi_{-\varkappa\mu} | \sigma_r | \chi_{\varkappa\mu})$ is obtained by using (5.20) which yields

$$(\tfrac{1}{2} \| \sigma \| \tfrac{1}{2}) = \sqrt{3}$$

and (5.19b) which now takes the form

$$(l' \| \mathbf{r} \| l) = -C(l'1l; 00)$$

This C-coefficient and the Racah coefficient in (6.21) are obtained from Tables I.2 and I.3, respectively. In (6.21) the following transcription of notation is needed: $j_1 j_2 j_1' j_2'; jL \rightarrow l\tfrac{1}{2}l'\tfrac{1}{2}; j1$. Finally, we find for both $l - l' = \pm 1$ that

$$\sigma_r \chi_{\varkappa\mu} = -\chi_{-\varkappa,\mu} \tag{9.12b}$$

Using these results, (9.12) becomes

$$
\left\{
\begin{array}{l}
-i\chi_{-\varkappa,\mu}\left[(W-V+1)f-g'-\dfrac{\varkappa+1}{r}g\right] \\[3mm]
\chi_{\varkappa\mu}\left[(W-V-1)g+f'-\dfrac{\varkappa-1}{r}f\right]
\end{array}
\right\} = 0 \qquad (9.13)
$$

Therefore each square bracket in (9.13) is zero, and these give the (real) radial wave equations for f and g.

It is not our purpose to study these radial equations. Rather, the point to be emphasized is that the assumed form (9.11) for Ψ is indeed a correct one and that this is a wave function in the angular momentum representation. The latter is now almost self-evident. However, to elaborate upon it we observe from the definition that

$$\mathbf{J}^2\chi_{\varkappa\mu} = j(j+1)\chi_{\varkappa\mu} \qquad (9.14a)$$

$$\mathbf{L}^2\chi_{\varkappa\mu} = l_\varkappa(l_\varkappa+1)\chi_{\varkappa\mu} \qquad (9.14b)$$

$$J_z\chi_{\varkappa\mu} = \mu\chi_{\varkappa\mu} \qquad (9.14c)$$

$$\mathbf{S}^2\chi_{\varkappa\mu} = \tfrac{3}{4}\chi_{\varkappa\mu} \qquad (9.14d)$$

Therefore, for the relativistic wave function (9.11)

$$\mathbf{J}^2\Psi = j(j+1)\Psi; \qquad \mathbf{J} = \mathbf{L} + \tfrac{1}{2}\boldsymbol{\sigma} \qquad (9.15)$$

because this is valid for both $\chi_{\pm\varkappa,\mu}$. Also

$$\mathbf{S}^2\Psi = \tfrac{3}{4}\Psi \qquad (9.15a)$$

and

$$J_z\Psi = \mu\Psi \qquad (9.15b)$$

However

$$\mathbf{L}^2\Psi = \begin{bmatrix} -i\,fl_{-\varkappa}(l_{-\varkappa}+1)\chi_{-\varkappa,\mu} \\[2mm] gl_\varkappa(l_\varkappa+1)\chi_{\varkappa\mu} \end{bmatrix}$$

and there is no value of $\varkappa$ for which $l_{-\varkappa}(l_{-\varkappa}+1) = l_\varkappa(l_\varkappa+1)$. In fact, the condition for this equality to hold is $1 + S(\varkappa)(2l_{-\varkappa}+1) = 0$ or $\varkappa < 0$ and $l_{-\varkappa} = 0$. The latter implies $\varkappa = 1$ so that the first condition is violated. Thus, $\mathbf{L}^2$ is not diagonal, which means that the orbital angular momentum is not a constant of the motion. In fact, $[\mathbf{L}^2, \boldsymbol{\alpha}\cdot\mathbf{p}] \neq 0$.

There is another operator which is diagonal. From (9.11) we see that

$$K\Psi = \beta(\boldsymbol{\sigma}\cdot\mathbf{L} + 1)\Psi = \varkappa\Psi \qquad (9.16)$$

In fact, since β and $\boldsymbol{\sigma}$ commute and $\beta^2 = 1$,

$$
\begin{aligned}
K^2 &= (\boldsymbol{\sigma}\cdot\mathbf{L} + 1)^2 = (\boldsymbol{\sigma}\cdot\mathbf{L})^2 + 2\,\boldsymbol{\sigma}\cdot\mathbf{L} + 1 \\
&= \mathbf{L}\cdot\mathbf{L} + i\,\boldsymbol{\sigma}\cdot(\mathbf{L}\times\mathbf{L}) + 2\,\boldsymbol{\sigma}\cdot\mathbf{L} + 1 \\
&= \mathbf{L}^2 + i\,\boldsymbol{\sigma}\cdot i\mathbf{L} + 2\,\boldsymbol{\sigma}\cdot\mathbf{L} + 1 \\
&= \mathbf{L}^2 + \boldsymbol{\sigma}\cdot\mathbf{L} + 1 \\
&= \mathbf{J}^2 + \tfrac{1}{4} \qquad (9.17)
\end{aligned}
$$

This result checks with the above since the eigenvalue of $\mathbf{J}^2 + \tfrac{1}{4}$ is $j(j+1) + \tfrac{1}{4} = (j + \tfrac{1}{2})^2 = \varkappa^2$.

We see that the relativistic motion of an electron is a sort of superposition of two kinds of Pauli motion: For a given j the wave function contains both $l = j + \tfrac{1}{2}$ and $l = j - \tfrac{1}{2}$, and this is just what is expected as arising from the spin-orbit coupling. Moreover, each of these l-values is associated with its own radial function (that is, f and g are linearly independent as the radial wave equations show). This is in contrast to the non-relativistic description.

Wave functions such as those described here are used extensively in the description of internal conversion, beta decay, electron scattering, and in many other problems. As an application we consider the shape of the energy spectrum in beta decay for the so-called S or scalar interaction, where reference is made to the transformation properties of the interaction under Lorentz transformations.

The interaction operator for β^- emission is

$$H_\beta = \sum_k \Psi_f^* \mathfrak{IC}_k \Psi_i (\psi^* \mathcal{L}\phi)_{\mathbf{r}_k} \qquad (9.18)$$

Here Ψ_f and Ψ_i are final- and initial-state nuclear wave functions while $\mathfrak{IC}_k$ is an operator in the space of the kth nucleon (located at $\mathbf{r}_k$),

$$\mathfrak{IC}_k = \beta_k \,\tau_-(k) \qquad (9.18a)$$

with $\tau_-(k)$ defined so that it gives zero when the kth particle is a proton and when acting on a neutron, as the kth particle, changes it into a proton; see section 39. This definition is given for completeness although the actual form of $\mathfrak{IC}_k$ is not essential for the spectrum shape. The operator $\mathcal{L}$, which is relevant, has the form

$$\mathcal{L} = \beta C \qquad (9.18b)$$

where $C = i\beta\alpha_y K_0$ where K_0, the complex conjugation operator, changes the neutrino wave function ϕ from the form (9.11) into one belonging to a negative energy state. H_β then destroys such a neutrino, creating an antineutrino of positive energy $q = W_0 - W$, where W is the electron energy and W_0 is the total energy release in the transition. The energy spectrum is then

$$N(W) = \frac{2W}{\pi p} \sum_{\varkappa\varkappa'} \sum_{\mu\mu'} \left| \int H_\beta \right|^2 \tag{9.19}$$

where $p = (W^2 - 1)^{\frac{1}{2}}$ is the electron momentum, $\varkappa\,\mu$ are the electron quantum numbers and $\varkappa'\,\mu'$ those of the neutrino.[4] The integral in (9.19) is over all nuclear coordinates, and this includes summations over nuclear substates of different projection quantum numbers. The light particle functions ψ and ϕ are evaluated at the position of the transforming nucleon, and the radial parts will eventually be evaluated at the nuclear radius ρ.

It is now seen that the quantity to be evaluated is

$$i\psi^*\alpha_y\phi = -f\,g_\nu(\chi_{-\varkappa\mu},\,\sigma_y\chi^*_{\varkappa'\mu'}) + g\,f_\nu(\chi_{\varkappa\mu},\,\sigma_y\chi^*_{-\varkappa'\mu'}) \tag{9.20}$$

In (9.20) the radial functions for the neutrino are designated by f_ν and g_ν and are,[4] for the normalization appropriate to (9.19),

$$f_\nu = S(\varkappa')q\,j_{\bar{l'}}(qr)$$

$$g_\nu = q\,j_{l'}(qr) \tag{9.21}$$

These are obtained from the radial equations, see (9.18), by setting $W - V \pm 1 = q$ and $\varkappa = \varkappa'$. Here and in the following we use $l = l_\varkappa$, $\bar{l} = l_{-\varkappa}$ and similarly for l' and $\bar{l}'$. To evaluate the scalar products in (9.20), which involve only spin sums, we note that

$$\sigma_y \psi_{\frac{1}{2}m} = i(-)^{m-\frac{1}{2}} \psi_{\frac{1}{2},-m}$$

so that

$$\sigma_y \chi^*_{\varkappa\mu} = i \sum_m (-)^{m-\frac{1}{2}} C(l\tfrac{1}{2}j;\,\mu-m,\,m)\,\psi_{\frac{1}{2},-m}\,Y^*_{l,\mu-m}$$

$$= i(-)^{\mu-\frac{1}{2}} \sum_m C(l\tfrac{1}{2}j;\,\mu-m,\,m)\,\psi_{\frac{1}{2},-m}\,Y_{l,m-\mu}$$

We change the sign of the summation index m and apply (3.16a) to the C-coefficient. Then

$$\sigma_y \chi^*_{\varkappa\mu} = i(-)^{\mu-\frac{1}{2}}(-)^{l+\frac{1}{2}-j} \sum_m C(l\tfrac{1}{2}j;\,-\mu-m,\,m)\,\psi_{\frac{1}{2}m}\,Y_{l,-m-\mu}$$

$$= i(-)^{l+\mu-j}\chi_{\varkappa,-\mu} \tag{9.22}$$

[4] M. E. Rose and R. Osborne, *Phys. Rev.* **93**, 1326 (1954).

The result (9.22) indicates that we need to evaluate the quantity

$$R_{\mu\mu'}(\varkappa\varkappa') = (\chi_{\varkappa\mu},\,\chi_{\varkappa'\mu'})$$

and then

$$i\,\psi^*\alpha_y\phi = i(-)^{l'+\mu'-j'}[R_{\mu,-\mu'}(-\varkappa,\varkappa')\,fg_\nu - R_{\mu,-\mu'}(\varkappa,-\varkappa')\,gf_\nu] \quad (9.23)$$

since $|\bar{l}' - l'| = 1$ and $j_{\varkappa'} = j_{-\varkappa'} = j'$.

From (9.4),

$$R_{\mu\mu'}(\varkappa\varkappa') = \sum_{mm'} C(l\tfrac{1}{2}j;\,\mu-m,\,m)\,C(l'\tfrac{1}{2}j';\,\mu'-m',\,m')\,\delta_{mm'}\,Y^*_{l,\mu-m}Y_{l',\mu'-m'}$$

$$= \sum_m (-)^{\mu-m}\,C(l\tfrac{1}{2}j;\,\mu-m,\,m)\,C(l'\tfrac{1}{2}j';\,\mu'-m,\,m)$$

$$\times \sum_\lambda \left[\frac{(2l+1)(2l'+1)}{4\pi(2\lambda+1)}\right]^{\frac{1}{2}} C(ll'\lambda;00)$$

$$\times\, C(ll'\lambda;\,m-\mu,\,\mu'-m)\,Y_{\lambda,\mu'-\mu} \quad (9.24)$$

by (4.32). The summation over m involving the phase $(-)^{\mu-m}$ and the three C-coefficients gives

$$\sum_m = (-)^{\mu+\frac{1}{2}}[(2j+1)(2j'+1)]^{\frac{1}{2}}\,C(jj'\lambda;\,-\mu,\,\mu')\,W(j'l'jl;\tfrac{1}{2}\lambda) \quad (9.25)$$

by a single Racah recoupling. We have used the fact that $l + l' + \lambda$ is an even integer to simplify a phase factor. Inserting (9.25) into (9.24) gives

$$R_{\mu\mu'}(\varkappa\varkappa') = \sum_\lambda (-)^{\mu+\frac{1}{2}}S_\lambda(\varkappa,\varkappa')\,C(jj'\lambda;\,-\mu,\mu')\,Y_{\lambda,\mu'-\mu} \quad (9.26a)$$

with

$$S_\lambda(\varkappa,\varkappa') = \left[\frac{(2l+1)(2l'+1)(2j+1)(2j'+1)}{4\pi(2\lambda+1)}\right]^{\frac{1}{2}}$$

$$\times\, C(ll'\lambda;00)\,W(j'l'jl;\tfrac{1}{2}\lambda) \quad (9.26b)$$

It will be observed that, since j and j' are independent of the signs of $\varkappa$ and $\varkappa'$, both terms in (9.23) have the factor $C(jj'\lambda;\,-\mu,\,-\mu')$. With this in mind we are now prepared to calculate the square of the matrix element in (9.19) and to perform the appropriate projection quantum number sums. Thus, with $\mu + \mu' = -M$,

$$\left|\int H_\beta\right|^2 \cong \left|\sum_\lambda C(jj'\lambda;\,-\mu,\,\mu+M)\right.$$

$$\times \{S_\lambda(-\varkappa,\varkappa')\,f\,g_\nu - S_\lambda(\varkappa,-\varkappa')\,g\,f_\nu\}_{r=\rho}\rho^{-\lambda}$$

$$\left.\times \int \Psi_f^* \sum_k \mathscr{K}_k \mathscr{Y}_{\lambda M}\Psi_i\right|^2 \quad (9.27)$$

$\mathcal{Y}_{\lambda M}$ is the usual solid harmonic, the radial factor is r_k^λ, the operator $\mathcal{K}$ is a zero-rank tensor, and therefore the operator in the integral in (9.27) is a tensor of rank λ. Accordingly, we may write it in the form

$$\int \Psi_f^* \sum_k \mathcal{K}\, \mathcal{Y}_{\lambda M}\, \Psi_i = C(J_i \lambda J_f; M_i, M_f - M_i)(f \parallel \cdots \parallel i)\, \delta_{M, M_f - M_i}$$

If we omit the reduced matrix element as an irrelevant multiplicative factor, the sums to be performed are

$$\sum_{MM_i} \sum_\mu \left| \sum_\lambda C(jj'\lambda; -\mu, M+\mu)\, C(J_i \lambda J_f; M_i M) \cdots \right|^2 \qquad (9.28)$$

where the dots indicate the remaining factors in (9.27) which depend on λ. The sum over μ' has now been replaced by a sum over M, and it is clear that the conservation rule

$$\mu + \mu' = -M = M_i - M_f$$

holds. This is just what was to be expected since $-M$ is the component on the quantization axis of the total angular momentum in the electron-neutrino field.

When the sum over λ in (9.28) is squared and the sum over μ carried out, with M held fixed, it is seen that cross-terms with different values of λ will give a vanishing contribution. The λth term is the usual λth order of forbiddenness contribution and these do not interfere because they are associated with tensors of different rank. Finally, the sum over M_i and M yields a factor $2J_f + 1$ according to

$$\sum_{MM_i} [C(J_i \lambda J_f; M_i M)]^2 = 2J_f + 1$$

The energy distribution of the β^- particles for the transition of forbiddenness order λ is

$$N_\lambda(W) \sim \frac{W}{p} \sum_{\varkappa \varkappa'} [S_\lambda(-\varkappa, \varkappa')\, f\, g_\nu - S_\lambda(\varkappa, -\varkappa')\, g\, f_\nu]^2{}_{r=\rho} \qquad (9.29)$$

The values of $\varkappa$ and $\varkappa'$ are those consistent with the triangular conditions $\Delta(j\, j'\, \lambda)$ and $\Delta(\bar{l}\, l'\, \lambda)$ or $\Delta(l\, \bar{l}'\, \lambda)$ with $\bar{l} + l' + \lambda$ or $l + \bar{l}' + \lambda$ an even integer; see (9.26b) and (9.27). The values of λ which are permitted are determined by the condition $\Delta(J_i J_f \lambda)$. It is in this way that the conservation of angular momentum is expressed. Because both $p\rho$ and $q\rho \ll 1$, the lowest value of λ consistent with the given values of J_i and J_f and parity change will make the most important contribution. The parity rule is

$$\pi_i \pi_f = (-)^\lambda$$

Thus for $\pi_i \pi_f = 1$ (no parity change), the lowest forbiddenness order is $\lambda = |J_i - J_f|$ or $|J_i - J_f| + 1$ whichever is even. The smallest values of $\varkappa$ and $\varkappa'$ consistent with the rules given above are then to be used in (9.29). For example, for $\lambda = 0$ only $\varkappa = -\varkappa' = \pm 1$ are needed.

The other interactions, including cross-terms, usually considered in the theory of beta decay can be treated in the same manner as the preceding and, of course, the energy distributions obtained are just the familiar ones which have been given elsewhere.[5]

The procedures illustrated in the foregoing have also been applied to the problem of internal conversion,[6] Auger effect, and photoelectric effect.[7] However, it should be emphasized that in these problems the value of radial matrix elements depending on the evaluation of radial functions at all points of space is involved. The computation of these matrix elements is very laborious and has been carried out only for the first-mentioned problem.

[5] E. J. Konopinski and G. E. Uhlenbeck, *Phys. Rev.* **60**, 308 (1941).

[6] M. E. Rose, *Beta and Gamma-Ray Spectroscopy*, Ed. K. Siegbahn, Chapter XIV, North Holland Publishing Co., Amsterdam, 1955.

[7] M. E. Rose and L. C. Biedenharn, *Oak Ridge Natl. Lab. Rept. 1779.*

X. ORIENTED NUCLEI AND ANGULAR CORRELATIONS

32. CAPTURE OF POLARIZED NEUTRONS BY POLARIZED NUCLEI

One of the interesting applications of angular momentum theory arises in connection with emission or absorption of radiation from states of a nucleus that are effectively oriented. By oriented we mean that the $2j+1$ substates (where j is the angular momentum of the nucleus) of the emitting (or absorbing) state do not all have the same population.[1] In this case the angular distribution of emitted radiation is no longer isotropic. In fact, the orientation process will always single out a direction in space so that space is no longer isotropic for the emitting nucleus. We shall discuss this question somewhat further in section 34. In this section we discuss a simple problem wherein not only the absorbing nucleus is oriented but this is also true of the absorbed particles. In this case it is not a change in angular distribution that is pertinent but a change in cross section itself. This change is brought about by the simultaneous orientation of absorber and absorbed particles, and random orientation of either destroys the effect. The discussion of this problem is included here because of its relation to the effect of orientation on the angular distribution.

In general, we consider two types of orientation. In one, alignment, the population of the substate with projection quantum number m is dependent on $|m|$ but not on its sign:

$$p(m) = p(-m) \tag{10.1}$$

There is clearly no magnetic moment in an aligned nucleus, but it could have an induced quadrupole moment: $Q \sim \{m^2 - \frac{1}{3}j(j+1)\}_{\text{avg}}$. It will be seen that in angular correlations (section 33) the intermediate nuclear state is aligned.

---

[1] Several methods whereby orientation may be achieved are discussed by A. Simon, M. E. Rose, and J. M. Jauch, *Phys. Rev.* **84**, 1155 (1951). The excellent discussions by R. J. Blin-Stoyle, M. A. Grace, and H. Halban and by S. R. de Groot and H. A. Tolhoek in Chapter XIX of *Beta and Gamma Ray Spectroscopy*, North Holland Publishing Co., should also be consulted. The work of Blin-Stoyle et al. contains a brief description of relevant experiments.

In the second type of orientation, referred to as polarization, the population $p(m)$ has a more general, character. In fact, it need have no symmetry and, in general, a necessary restriction is that $p(m)$, as a function of m, must have an odd part so that (10.1) no longer applies. A nonvanishing value of m_{avg} and a magnetic moment is then to be expected. The case of a nuclear state of spin $\frac{1}{2}$ is interesting. In this case there are only two substates: $m = \pm\frac{1}{2}$. Then

$$p(\tfrac{1}{2}) = p_0 + \Delta p$$

$$p(-\tfrac{1}{2}) = p_0 - \Delta p \tag{10.2}$$

where p_0 is the population for no polarization and Δp is the change produced by the polarization process. Thus $p(m)$ is composed of two parts, one independent of m and an additional part which is odd in m. Now all the effects discussed, anisotropy of angular distributions, changes in cross sections, are linear in $p(m)$. We are interested only in deviations from uniform population, and so only $p(m) - p_0$ is of interest. For the spin $\frac{1}{2}$ case, (10.2) shows that $p(m) - p_0$ is an odd function of m. It is seen, then, by this simple argument, that alignment is impossible for states with $j = \frac{1}{2}$. In angular correlation one always gets isotropy if an intermediate state is characterized by $j = \frac{1}{2}$.

The change in absorption cross section which occurs when polarized nuclei absorb polarized neutrons is not subject to any such limitation. Consider the case of slow (s) neutrons, and let j' and j refer to the compound and target states, respectively. Then

$$j' = j \pm \tfrac{1}{2}$$

Let m', m, and μ be eigenvalues of j'_z, j_z, and i_z, where $\mathbf{i}$ is the neutron spin operator. Then $\mu = \pm\frac{1}{2}$, and weights $|C_{\pm}|^2$ are assigned to these neutron spin states. The neutron polarization is

$$f_n = |C_+|^2 - |C_-|^2 \tag{10.3}$$

where

$$|C_+|^2 + |C_-|^2 = 1$$

is assumed. Thus $-1 \le f_n \le 1$, and the equality signs mean complete polarization. The quantity f_n can be interpreted as the ratio of the net neutron moment to the saturation value. Let $p(m)$ be the population in the mth state of the target nucleus, and

$$p(m, \mu) = |C_{\pm}|^2 \, p(m) \tag{10.4}$$

will be the population parameters for the combined system of target nucleus and neutrons. Let Φ, χ, and ψ be state vectors of the compound state, neutron, and target nucleus, respectively. Then the cross section for absorption is

$$\sigma = \text{const} \sum_{m\mu} p(m, \mu) \left| (\Phi_{j'm'}, \chi_{\frac{1}{2}\mu}\, \psi_{jm}) \right|^2$$

$$= \text{const} \sum_{m\mu} p(m, \mu)\, C(j\tfrac{1}{2}j'; m\mu)^2 \tag{10.5}$$

This follows from the fact that $\mathbf{j'} = \mathbf{j} + \mathbf{i}$.

At this juncture it can be verified that for uniformly populated states the cross section is independent of j' except insofar as the constant factor in (10.5) may depend on j'. With the non-uniform weighting, which the polarization entails, the sum in (10.5) changes and *is* dependent on j'. It is also evident that the polarization process cannot change the total number of absorption events but can redistribute them between the two possible j'-values. We now evaluate the cross section for $j' = j \pm \frac{1}{2}$. Consider $j' = j + \frac{1}{2}$. Then, from Table I.1,

$$\sigma_+ = \text{const} \sum_{m\mu} p(m, \mu)\, \frac{j \pm m' + \frac{1}{2}}{2j + 1}$$

where the upper sign is for $\mu = \frac{1}{2}$, the lower for $\mu = -\frac{1}{2}$. This gives

$$\sigma_+ = \text{const} \sum_{m} p(m) \left[|C_+|^2 \frac{j + m + 1}{2j + 1} + |C_-|^2 \frac{j - m + 1}{2j + 1} \right]$$

We now introduce the nuclear polarization which is defined in analogy with (10.3) and has a similar physical interpretation.

$$f_N = \frac{1}{j} \sum_{m} m\, p(m)$$

As a matter of convenience we normalize according to

$$\sum_{m} p(m) = 1$$

Then

$$\sigma_+ = \text{const} \sum_{m} \left[\frac{j + 1}{2j + 1}\, p(m) + \frac{m\, p(m)}{2j + 1} f_n \right]$$

$$= \text{const} \left[\frac{j + 1}{2j + 1} + \frac{j f_n f_N}{2j + 1} \right]$$

Normalizing the constant factor in σ_+ so that for $f_n f_N = 0$ the cross section is σ_0 (that pertaining to no polarization of target nuclei and/or neutrons), we see that

$$\sigma_+ = \sigma_0 \left(1 + \frac{j}{j+1} f_n f_N \right) \qquad (10.6)$$

When $j' = j - \frac{1}{2}$

$$\sigma = \sigma_- = \text{const} \sum_{m\mu} p(m, \mu) \frac{j \mp m' + \frac{1}{2}}{2j+1}$$

Here the upper (lower) sign are for $\mu = +\frac{1}{2}$ ($-\frac{1}{2}$). Then

$$\sigma_- = \text{const} \sum_m p(m) \left[|C_+|^2 \frac{j - m}{2j+1} + |C_-|^2 \frac{j+m}{2j+1} \right]$$

$$= \text{const} \left[\frac{j}{2j+1} - \frac{j f_n f_N}{2j+1} \right] = \sigma_0 (1 - f_n f_N) \qquad (10.7)$$

The σ_0 in (10.6) and (10.7) are, of course, not the same since they refer to different compound states. In any given experiment we deal with one or the other of these two cases—provided the effects of the resonances can be separated. The point is that, for a given relative orientation of target nuclei and neutrons, the change of cross section is opposite in the two cases. The opposite signs of the cross section change was explained above. The results (10.6) and (10.7) can be understood in a fairly simple way. First $\Delta\sigma/\sigma_0$ should be proportional to $f_n f_N$. Second for $j' = j - \frac{1}{2}$ the case $f_n f_N = 1$ should give $\sigma_- = 0$ since the only substate that can be populated is $m' = \pm(m + \frac{1}{2})$, and this substate does not exist for $j' = j - \frac{1}{2}$. This makes the result (10.7) evident. Then (10.6) follows from the fact that, with σ_0 assumed to be the same for the two compound states, we would have $\sigma_{\text{avg}} = \sigma_0$ where the average is one taken with the weight factors $2j' + 1$.

If we know the sign of the polarization and j', the direction of the cross-section change can be predicted. Conversely, j' can be determined, if the sign of $f_n f_N$ is known, by observing the cross section with and without polarization.

There are two methods of polarization commonly used. The first, referred to as the "brute force" or direct coupling method, utilizes a magnetic field directly applied to the nuclear moment. Hence, the sign of the nuclear g-factor must be known. In the second method the magnetic field produced by the hyperfine coupling in a paramagnetic salt is utilized so that we need to know the sign of the relevant coupling constant. This is often available from microwave data. In this second method only a small field is applied to polarize the electrons and the contribution of the brute force effect is small.

33. ANGULAR CORRELATION

The literature of angular correlation theory is very complete. Many comprehensive and rigorous treatments of the subject have been given.[2] Here we do not intend to be rigorous, nor do we attempt to present the theory in its most elegant form. What will be attempted is a simple description which should be easy to follow in terms of the principles already developed.

Briefly the phenomenon involves the cascade emission of two or more radiations. If two of these are observed in coincidence, only the directions of propagation being measured, we speak of a direction–direction correlation. The coincidence counting rate will then be considered as a function of the angle between the propagation vectors of the two radiations. In this simplest case we suppose that the two transitions are consecutive. Generalizations are possible in which unobserved radiations intervene and in which the polarization of one of the radiations is also measured. These and other complications, arising from the perturbation of the nucleus in its intermediate state, will not concern us.

First, it is essential to understand why there is a correlation; that is, why does the coincidence counting rate depend on θ, the afore-mentioned angle between the propagation vectors? Consider a transition in which some type of radiation is emitted from a state with quantum numbers j, m leading to a state j', m'. The radiation will have angular momentum L and projection quantum number $M = m - m'$. Although several L-values ranging from $|j - j'|$ to $j + j'$ may be possible, this multiplicity of L-values is of no importance for the present argument, and we will continue the discussion in terms of a single value of L. In actual cases, of course, one observes transitions from all substates m to all substates m'. Although the angular distribution of the radiation emitted is, for a particular M, dependent on θ, the summation over all substates, which is equivalent to a summation over M, is independent [3] of θ. Hence the radiation observed in a single transition, from a non-oriented source, is isotropic as expected.

However, the situation is somewhat different for the coincidence observation of two radiations. If we consider a fixed M, the radiation is not isotropic, as mentioned, because a particular direction in space is

[2] For example, G. Racah, *Phys. Rev.* **84**, 910 (1951); L. C. Biedenharn and M. E. Rose, *Revs. Mod. Phys.* **25**, 729 (1953); F. Coester and J. M. Jauch, *Helv. Phys. Acta* **XXVI**, 3 (1953). A comprehensive but more elementary review is given by H. Frauenfelder, *Beta and Gamma Ray Spectroscopy*, North Holland Publishing Co., Chapter XIX, 1955.

[3] In the formalism this is a consequence of the unitary property of the matrix D^L; see equation (7.38).

being singled out. This is the quantization axis with reference to which M, as well as m and m', is measured. If, of all the radiation, emitted in all directions, we observe only that which proceeds in a specified direction, then this is equivalent to selecting a particular set of values of M. For instance, if we choose the quantization axis along the direction of observation, $M = 0$ for spinless particles and $M = \pm\frac{1}{2}$ for particles of intrinsic spin $\frac{1}{2}$ while for photons $M = \pm 1$; the explanation of these statements is given below. As a consequence, we do not sum over all values of M, and in the final state not all values of m' are equally populated. If now this final state for the first transition is also the initial state for the next transition, it is clear that the ensuing radiation is, in general, anisotropic. Because both positive and negative values of M are selected with equal weight, it will appear that the type of orientation that occurs is an alignment.

To conclude this preliminary discussion we consider the problem of a particle with a specified propagation vector $\mathbf{f}$ and investigate the possible components of the angular momentum about the direction of this vector. For a particle without spin the wave function is [4]

$$e^{i\mathbf{k}\cdot\mathbf{r}} = 4\pi \sum_{l=0}^{\infty} \sum_{m=-l}^{l} i^l j_l(kr) \, Y_{lm}^*(\mathbf{f}) \, Y_{lm}(\mathbf{r}) \tag{10.8}$$

When $\mathbf{f}$ is along the axis of quantization, the polar angle in $Y_{lm}^*(\mathbf{f})$ is zero and $Y_{lm}^* = 0$ unless $m = 0$. This is true for all l, and it does not matter whether in a given transition one or more than one term in the plane wave sum contributes to the matrix elements.

If the particle has intrinsic spin s, the wave function [4] is $\chi_{s,m_s} e^{i\mathbf{k}\cdot\mathbf{r}}$. Inverting the product $\chi_{s,m_s} Y_{lm}(\mathbf{r})$ according to (3.8), we find

$$\chi_{s,m_s} e^{i\mathbf{k}\cdot\mathbf{r}} = 4\pi \sum_{l=0}^{\infty} \sum_{m=-l}^{l} \sum_{j} i^l j_l(kr) \, Y_{lm}^*(\mathbf{f}) \, C(lsj;\, mm_s) \, \psi_{j,m+m_s} \tag{10.9}$$

which is an expansion in eigenfunctions of the total angular momentum. When $\mathbf{f}$ is along the quantization axis, $m = 0$ and $M = m + m_s = m_s$. For neutrons, protons, and for light fermions treated non-relativistically, $M = \pm\frac{1}{2}$.

---

[4] Strictly speaking, this is the wave function only if the particle is assumed to move in a field-free region. If this is not the case, an appropriate wave function is obtained by changing the radial function and inserting, in the lth term on the right-hand side of (10.8), the phase factor $\exp(-i\sigma_l')$ where σ_l' is the phase shift for the lth partial wave in the external field; see G. Breit and H. A. Bethe, *Phys. Rev.* **93**, 888 (1954). For the Coulomb field σ_l' is given as the second term in (10.53) below. It will be evident that the conclusions to be drawn below are in no way restricted by the assumption of a free particle.

That the same conclusion applies to relativistic fermions is also to be expected. To see this in detail, we write a plane wave for a relativistic particle in the form

$$\psi = A \, e^{i\mathbf{p}\cdot\mathbf{r}}$$

where $\mathbf{p}$ is the momentum and the amplitude A is a four-component spinor. Writing it in terms of two Pauli (two-component) spinors,

$$A = \begin{pmatrix} u \\ v \end{pmatrix}$$

we find, from (9.7) with $V = 0$,

$$(W + 1)u = -\boldsymbol{\sigma}\cdot\mathbf{p}v$$

$$(W - 1)v = -\boldsymbol{\sigma}\cdot\mathbf{p}u$$

A complete set of (positive energy) plane waves is obtained by setting

$$u = \psi_{\frac{1}{2}m}, \qquad m = \pm\tfrac{1}{2}$$

That is, u is one or the other of the Pauli spin functions. Expanding the plane wave in eigenfunctions of the angular momentum, see (9.11), we get

$$A \, e^{i\mathbf{p}\cdot\mathbf{r}} = \sum_j C_{j\mu} \Psi_{j\mu} \tag{10.8a}$$

where, for greater clarity, we have added the indices j,μ to Ψ. Here μ has the same meaning as M in the foregoing. Using (9.11), we may write (10.8a) in the form

$$\psi_{\frac{1}{2}m} \, e^{i\mathbf{p}\cdot\mathbf{r}} = \sum_{\varkappa\mu} C_{\varkappa\mu} \, g_\varkappa \chi_{\varkappa\mu} \tag{10.8b}$$

$$\frac{\boldsymbol{\sigma}\cdot\mathbf{p}}{W + 1} \psi_{\frac{1}{2}m} \, e^{i\mathbf{p}\cdot\mathbf{r}} = i \sum_{\varkappa\mu} C_{\varkappa\mu} \, f_\varkappa \chi_{-\varkappa\mu} \tag{10.8b'}$$

Taking the scalar product with $\chi_{\varkappa\mu}$, on the surface of a sphere of radius r, of the first equation, we obtain

$$g_\varkappa \, C_{\varkappa\mu} = (\chi_{\varkappa\mu}, \psi_{\frac{1}{2}m} \, e^{i\mathbf{p}\cdot\mathbf{r}}) \tag{10.8c}$$

The second equation (10.8b') gives equivalent results, as is seen when use is made of the differential equations defining f and g; see (9.13).

Expanding the plane wave as in (10.8) and using $g_\varkappa = j_l(kr)$ where $l = l_\varkappa$, we obtain

$$j_l \, C_{\varkappa\mu} = 4\pi \sum_\tau C(l\tfrac{1}{2}j; \mu-\tau, \tau)(\psi_{\frac{1}{2}\tau}, \psi_{\frac{1}{2}m}) \sum_{\lambda m'} i^\lambda j_\lambda \, Y^*_{\lambda m'}(\mathbf{p}) \, (Y_{l,\mu-\tau}, Y_{\lambda m'})$$

The scalar products give $\delta_{\tau m}$ and $\delta_{\mu-m,m'}$ as well as $\delta_{l\lambda}$. Hence we obtain

$$C_{\varkappa\mu} = 4\pi i^l \, C(l\tfrac{1}{2}j, \mu-m, m) \, Y_{l,\mu-m}(\mathbf{p}) \tag{10.8d}$$

For $\mathbf{p}$ along the z-axis the index $\mu - m$ in $Y_{l,\mu-m}$ must be zero or $\mu = m = \pm\frac{1}{2}$. This is the required result.

Finally we consider the case of a plane electromagnetic wave. The procedure is analogous to the foregoing. A plane wave of arbitrary polarization is a linear combination of the circularly polarized plane waves considered in section 27. The expansion into multipole fields was carried out there, and, as equation (7.37) shows, the only possible values for the projection of the angular momentum along the axis of propagation are $p = \pm 1$.

We are now prepared to derive the form of the correlation function (proportional to the uncorrected coincidence counting rate) as a function of θ for the double cascade correlation. That is, two consecutive radiations are considered. We make explicit use of perturbation theory only to make the discussion as simple as possible. The only aspect of it that enters is the dependence of matrix elements on the projection quantum numbers, and this is given quite generally by the Wigner–Eckart theorem. Therefore the question of validity of the perturbation theory is irrelevant.

For the system of nucleus [5] plus field of emitted radiation, we need only consider the Hamiltonian H_0 of the nucleus alone and the interaction energy H_r responsible for emission or absorption. Then for the state vector Ψ we have

$$i\,\dot{\Psi} = (H_0 + H_r)\,\Psi \tag{10.10}$$

where we have set $\hbar = 1$ and the dot corresponds to time differentiation. It will suffice to discuss three states labeled eventually by j,m-values. For the moment let the amplitudes of these states be called $a_1(t)$, $a_j(t)$ and $a_2(t)$ for the initial, intermediate, and final states, respectively. Also ψ_1, ψ_j, and ψ_2 represent stationary state wave functions for these states. Then we set

$$\Psi = \psi_1\,a_1(t)\,e^{-iE_1 t} + \psi_j\,a_j(t)\,e^{-iE_j(t)} + \psi_2\,a_2(t)\,e^{-iE_2 t} \tag{10.11a}$$

where

$$H_0\,\psi_1 = E_1\,\psi_1 \tag{10.11b}$$

and similarly for ψ_j and ψ_2 with (nuclear) energy values E_j and E_2.

Substitution of (10.11a) into (10.10) and use of (10.11b) gives, as usual,

$$i\,\dot{a}_1 = (\psi_1|H_r|\psi_j)\,e^{i\omega_1 t}\,a_j \tag{10.11c}$$

$$i\,\dot{a}_j = (\psi_j|H_r|\psi_1)\,e^{-i\omega_1 t}\,a_1 + (\psi_j|H_r|\psi_2)\,e^{i\omega_2 t}\,a_2 \tag{10.11d}$$

$$i\,\dot{a}_2 = (\psi_2|H_r|\omega_j)\,e^{-i\omega_2 t}\,a_j \tag{10.11e}$$

[5] It is not necessary that the source of radiation be a nucleus although in most practical applications this is the case.

We have introduced the notation

$$\omega_1 = E_1 - E_j, \qquad \omega_2 = E_j - E_2$$

The initial conditions

$$a_1(0) = 1, \qquad a_j(0) = a_2(0) = 0 \qquad (10.11f)$$

are to be applied to the solution of (10.11c–10.11e). Then these initial conditions are inserted in (10.11d) to yield

$$a_j = (\psi_j|H_r|\psi_1)\frac{e^{-i\omega_1 t} - 1}{\omega_1}$$

so that (10.11e) becomes

$$i\,\dot{a}_2 = (\psi_2|H_r|\psi_j)(\psi_j|H_r|\psi_1)\frac{e^{-i(\omega_1+\omega_2)t} - e^{-i\omega_2 t}}{\omega_1}$$

Integration over t yields

$$a_2 = (\psi_2|H_r|\psi_j)(\psi_j|H_r|\psi_1)\frac{1}{\omega_1}\left[\frac{e^{-i(\omega_1+\omega_2)t} - 1}{\omega_1 + \omega_2} - \frac{e^{-i\omega_2 t} - 1}{\omega_2}\right]$$

when the integration constant has been adjusted to conform to $a_2(0) = 0$.

In lowest order a_2, the final state probability amplitude, contains one matrix element for each of the two transitions. The operator H_r depends on the properties of the radiation emitted so that this operator is not exactly the same in the two matrix elements, contrary to the impression that might be created by the abbreviated notation. Thus, a_2 is the probability amplitude that the nucleus be in the final state having emitted radiations of specified character. This includes, of course, a specification of propagation directions. Hence we are interested in $|a_2|^2$ summed over all final energy states. This summation converts the time-dependent factor into $2\pi t$, and the transition rate, $d\sum|a_2|^2/dt$, is proportional to

$$W = \sum\left|(2|H_r|j)(j|H_r|1)\right|^2 \qquad (10.12)$$

The summation is over substates of the final state, and an average over initial substates is also implied.

Strictly speaking there are $2j+1$ intermediate states, and the cascade may proceed via any one of them. Consequently we should write

$$W = \frac{1}{2j_1 + 1}\sum_{m_1 m_2}\left|\sum_m (j_2 m_2|H_r|jm)(jm|H_r|j_1 m_1)\right|^2 \qquad (10.13)$$

However, we can simplify this result by choosing the quantization axis along the propagation direction of either the first or second radiation. For definiteness let it be chosen along the direction of radiation 1. Then M_1, the projection quantum number of radiation 1, has either one of two possible values, depending on the nature and polarization state (if any) of the radiation. Whatever the number of possible values of M_1, we either select one of these if we are concerned with polarized radiation or average, or equivalently, sum over them incoherently, if the polarization of the radiation is not observed. This is because unpolarized radiation is a superposition of polarized radiations with completely random relative phases. The superposition is therefore an incoherent mixture. Hence, for a given m_1 we sum over m incoherently because $m_1 = M_1 + m$, and m_1 is also summed incoherently. Consequently the cross-terms in the coherent sum in (10.13) are unnecessary and

$$W = S \sum_{m_1 m_2 m} |(j_2 m_2 | H_r | jm)(jm | H_r | j_1 m_1)|^2 \qquad (10.14)$$

where a constant factor has been omitted. This is just the same as (10.12). Of course, it is again possible to be more general and keep the quantization axis arbitrary. The result, as expected, is just the same as is obtained from (10.14). The summation in (10.14) should also include summing over all non-observed quantities—for instance, the polarization quantum numbers if the experiment does not involve polarization measurements. To indicate this summation we have introduced the symbol S in (10.14) as an explicit reminder.

It is most convenient to break up the correlation function into two parts in the following way: We define the matrix

$$\Lambda_{mm'}^{(1)} = S_1 \sum_{m_1} (jm | H_r | j_1 m_1)(jm' | H_r | j_1 m_1)^* \qquad (10.15)$$

which is characteristic of the first transition, and the matrix

$$\Lambda_{m'm}^{(2)} = S_2 \sum_{m_2} (j_2 m_2 | H_r | jm)(j_2 m_2 | H_r | jm')^* \qquad (10.16)$$

and this latter matrix plays the same role for the second transition as $\Lambda_{mm'}^{(1)}$ does for the first.[6] Here S_1 and S_2 refer, respectively, to summa-

[6] In fact if we use the Hermitian property

$$(j_2 m_2 | H_r | jm) = (jm | H_r | j_2 m_2)^*$$

then $\Lambda^{(2)}$ is the same function of $j_2 m_2 \alpha_2$ as $\Lambda^{(1)}$ is of $j_1 m_1 \alpha_1$. The α_2 and α_1 refer to the parameters descriptive of the radiations emitted in second and first transitions. The Λ-matrices are actually density matrices; see L. C. Biedenharn and M. E. Rose, *Revs. Mod. Phys.* **25**, 729 (1953).

tions over unobserved variables of first and second radiations (i.e., $S = S_1 S_2$).

It is clear now that the correlation function is

$$W = \sum_m \Lambda^{(1)}_{mm} \Lambda^{(2)}_{mm} \qquad (10.17a)$$

If we had not chosen the axis of quantization along the direction of propagation of one of the radiations, we would have instead

$$W = \sum_{mm'} \Lambda^{(1)}_{mm'} \Lambda^{(2)}_{m'm} = \text{trace } \Lambda^{(1)} \Lambda^{(2)} \qquad (10.17b)$$

The choice corresponding to (10.14) diagonalizes the Λ-matrices. In any event the correlation function is given by the trace of $\Lambda^{(1)} \Lambda^{(2)}$, and this is a statement that is independent of the representation.

The advantage of breaking up the correlation function in this manner is that each transition is studied separately. Thus, after one obtains a γ–γ correlation function say, it is a much simpler matter to study the p–γ process where a proton is captured. The matrix $\Lambda^{(2)}$ will then be of the same form, and only the $\Lambda^{(1)}$-matrix need be changed. It will be seen that these matrices have a general form which facilitates the investigation of any correlation function in terms of the general result.

To determine the form of the Λ-matrices and, hence the general correlation function, we observe that for each radiation the interaction energy H_r is an invariant and must have the form [see (5.3)],

$$H_r = \sum_L b_L \sum_{M=-L}^{L} (-)^M A_{L,-M} T_{LM} \qquad (10.18)$$

Here T_{LM} is a tensor of rank L in the coordinates of the nucleus. For example, for gamma emission it could be the multipole moments derived from the charge and current densities in the nucleus.[7] The tensor components A_{LM} (also of rank L) refer to the radiated field. For gamma rays they would be obtained from the vector potential of the plane wave (7.37) when T_{LM} is obtained from the charge and current densities. The b_L are certain constants which we need not inquire about at this juncture because they are actually irrelevant for our purpose, and they can be lumped into the $A_{L,-M}$ or the T_{ML}. It is possible to assign a simple

[7] Actually the precise nature of the operator T_{LM} is not known, but this is essentially irrelevant for our purposes. Its properties as an irreducible tensor are all that will be of concern. This is the same point that has been emphasized on several occasions: that the information required is contained in the symmetry of the problem as evidenced by the transformation properties of relevant tensors. The physics as expressed in terms of some model, plays a role in affecting the value of weight constants when tensors of different rank are operative; see end of sections 33 and 34.

interpretation to these constants (multiplied by reduced matrix elements of T_L) in terms of relative intensities of emitted radiations of different angular momenta and this will be demonstrated in the following. Finally, if we choose the quantization axis along the propagation direction, M will have a much more restricted range of values, as discussed above.

Considering the case of radiation emitted in an arbitrary direction, we make a rotation of the quantization axis from the direction $\mathfrak{k}_1$ to some direction $\mathfrak{n}$ making an arbitrary angle with $\mathfrak{k}_1$. The tensors in the two reference frames are related by

$$T'_{LM} = \sum_\mu T_{L\mu} D^L_{\mu M}(R_1) \tag{10.19}$$

where R_1 collectively denotes the Euler angles of this rotation. Then the $\Lambda^{(1)}$ matrix becomes

$$\Lambda^{(1)}_{mm'} = S_1 \sum_{m_1} \sum_{LL'} \sum_{MM'} \sum_{\mu\mu'} (-)^{M+M'}$$

$$\times A_{L,-M} A^*_{L',-M'} D^L_{\mu M} D^{L'*}_{\mu'M'}(jm\,|\,T_{L\mu}\,|\,j_1m_1)(jm\,|\,T_{L'\mu'}\,|\,j_1m_1)^*$$

and we have redefined $A_{L,-M}$ so as to include the constants b_L in (10.18). We use (4.22) and the Wigner–Eckart theorem which permits us to write

$$\Lambda^{(1)}_{mm'} = S_1 \sum_{m_1} \sum_{LL'} \sum_{MM'} \sum_{\mu\mu'} (-)^{M-\mu'}(j\,\|\,T_L\,\|\,j_1)(j\,\|\,T_{L'}\,\|\,j_1)^*$$

$$\times A_{L,-M} A^*_{L,-M'} D^L_{\mu M} D^{L'}_{-\mu',-M'} C(j_1Lj;\,m_1\mu)$$

$$\times C(j_1L'j;\,m_1\mu')\,\delta_{m_1+\mu,m}\,\delta_{m_1+\mu',m'}$$

Replacing the product of the two D-matrices by the equivalent expression from the Clebsch–Gordan series, see (4.25), gives

$$\Lambda^{(1)}_{mm'} = S_1 \sum_{m_1} \sum_{LL'} \sum_{MM'} (-)^{M-m_1+m'}(j\,\|\,T_L\,\|\,j_1)(j\,\|\,T_{L'}\,\|\,j_1)^*$$

$$\times A_{L,-M} A^*_{L,-M'} C(j_1Lj;\,m_1,\,m-M_1)\,C(j_1L'j;\,m'_1,\,m'-M_1)$$

$$\times \sum_\nu C(LL'\nu;\,m-m_1,\,m_1-m')\,C(LL'\nu;\,M,\,-M')\,D^\nu_{m-m',M-M'}(R_1) \tag{10.20}$$

where the sums over μ and μ' have been done as well.

In (10.20) there is a sum over m_1 which can be carried out at once. To do this use (3.16a) and (3.16b) to write

$$C(j_1Lj;\,m_1,\,m-m_1) = (-)^{j_1-m_1}(-)^{j_1+j-L}\left[\frac{2j+1}{2L+1}\right]^{\frac{1}{2}}C(j_1jL;\,-m_1m)$$

$$= (-)^{j_1-m_1}\left[\frac{2j+1}{2L+1}\right]^{\frac{1}{2}}C(jj_1L;\,m,\,-m_1)$$

Combining with the first C-coefficient in the ν sum, we have

$$C(j_1Lj; m_1, m-m_1) \, C(LL'\nu; m-m_1, m_1-m')$$

$$= (-)^{j_1-m_1}(2j+1)^{\frac{1}{2}} \sum_t (2t+1)^{\frac{1}{2}} \, C(j_1L't; -m_1, m_1-m')$$

$$\times \, C(jt\nu; m, -m') \, W(jj_1\nu L'; Lt)$$

Therefore the sum over m_1 is

$$\sum_{m_1} (-)^{M-m_1+m'} \, C(j_1Lj; m_1, m-m_1) \, C(j_1L'j; m_1', m'-m_1)$$

$$\times \, C(LL'\nu; m-m_1, m_1-m')$$

$$= (-)^{M+m'+j_1}(2j+1)^{\frac{1}{2}} \sum_t (2t+1)^{\frac{1}{2}} \, W(jj_1\nu L'; Lt) \, C(jt\nu; m, -m')$$

$$\times \sum_{m_1} C(j_1L't; -m_1, m_1-m') \, C(j_1L'j; m_1', m'-m_1)$$

$$= (-)^{M+m'-L'+j}(2j+1) \, W(jj_1\nu L'; Lj) \, C(jj\nu; m, -m')$$

$$= (-)^{M+m'-L-L'+\nu+j_1}(2j+1) \, W(jjL'L; \nu j_1) \, C(jj\nu; m, -m')$$

$$(10.21)$$

Again $(6.4b)$ is used to obtain the first equality, and, after application of $(3.16a)$ in the m_1 sum, (3.7) is used. The last line results from $(6.10b)$ which reorders the arguments in the Racah coefficient.

We now set $M' = M + \tau$ and define the parameters

$$c_{\nu\tau}(LL') = S_1 \sum_M (-)^{L-M} A_{L,-M} A^*_{L',-M-\tau} \, C(LL'\nu; M, -M-\tau)$$

$$(10.22)$$

These parameters are characteristic of the radiation emitted and are independent of the properties of the nuclear states involved in the transition. In terms of these $c_{\nu\tau}$, we have

$$\Lambda^{(1)}_{mm'} = (2j+1) \sum_{LL'} \sum_{\nu\tau} (-)^{m'-L'+j_1+\nu}(j \parallel T_L \parallel j_1)(j \parallel T_{L'} \parallel j_1)^* c_{\nu\tau}$$

$$\times \, W(jjL'L; \nu j_1) \, C(jj\nu; m, -m') \, D^\nu_{m-m',-\tau}(R_1) \quad (10.23)$$

With regard to the parameter τ it is readily seen that this must be zero where only the propagation direction is observed; $\tau \neq 0$ refers to a polarization observation. This follows when it is remembered that the rotation denoted by R_1 depends only on two angles needed to specify a direction. Only when the orientation of a complete reference frame (f and a polarization direction) is needed, will the three Euler angles enter. The absence of any dependence on the third Euler angle means $\tau = 0$. This also implies that $M = M'$. Since M designates the polarization state, it means that for unpolarized radiation we sum over polarization states incoherently, as expected.

Therefore, for the directional correlation

$$\Lambda_{mm'}^{(1)} = (2j + 1) \sum_{LL'} \sum_{\nu} (j \parallel T_L \parallel j_1)(j \parallel T_{L'} \parallel j_1)^*(-)^{m'-L'+j_1+\nu} c_{\nu 0}$$

$$\times W(jjL'L; \nu j_1) \, C(jj\nu; m, -m') \, D_{m-m',0}^{\nu}(R_1) \quad (10.24)$$

In either (10.23) or (10.24) it is to be noted that $\Lambda_{mm'}^{(1)}$ depends on the projection quantum numbers only through $(-)^{m'} C(jj\nu; m, -m') D_{m-m',0}^{\nu}$. Thus, we may write

$$\Lambda_{mm'}^{(1)} = \sum_{\nu} (-)^{m'-j} C(jj\nu; m, -m') \, D_{m-m',0}^{\nu}(R_1) \, B_{\nu}(1) \quad (10.24a)$$

and

$$\Lambda_{m'm}^{(2)} = \sum_{\nu'} (-)^{m-j} C(jj\nu'; m, -m') \, D_{m'-m,0}^{\nu'}(R_2) \, B_{\nu'}(2) \quad (10.24b)$$

The parameters B_{ν} do not depend on projection quantum numbers. Of course, R_2 is the rotation which takes $\mathfrak{f}_2$ into $\mathfrak{n}$. Then the correlation function is

$$W = \sum_{mm'} \Lambda_{mm'}^{(1)} \Lambda_{m'm}^{(2)}$$

$$= \sum_{\nu\nu'} \sum_{m} \sum_{s} (-)^s \, B_{\nu}(1) \, B_{\nu'}(2) \, C(jj\nu; m, s-m) \, C(jj\nu'; m, s-m)$$

$$\times D_{s0}^{\nu}(R_1) \, D_{-s,0}^{\nu'}(R_2)$$

where $s = m - m'$. The sum over m with s fixed is obvious and gives $\delta_{\nu\nu'}$. Then

$$W = \sum_{\nu} \sum_{s} (-)^s \, B_{\nu}(1) \, B_{\nu}(2) \, D_{s0}^{\nu}(R_1) \, D_{-s0}^{\nu}(R_2)$$

But, because the D-matrices are unitary,

$$D_{-s0}^{\nu}(R_2) = D_{0,-s}^{\nu}{}^*(R_2^{-1}) = (-)^{-s} D_{0s}^{\nu}(R_2^{-1})$$

Hence the s summation is

$$\sum_{s} D_{0s}^{\nu}(R_2^{-1}) \, D_{s0}^{\nu}(R_1) = D_{00}^{\nu}(R_2^{-1}R_1) \quad (10.25a)$$

by the rule of matrix multiplication. But $R_2^{-1}R_1$ is the rotation which first carries $\mathfrak{f}_1$ into $\mathfrak{n}$ and then $\mathfrak{n}$ into $\mathfrak{f}_2$; i.e., it carries $\mathfrak{f}_1$ into $\mathfrak{f}_2$ independent of the arbitrary direction $\mathfrak{n}$. This shows that, as expected, the correlation depends only on $\cos \theta = \mathbf{f}_1 \cdot \mathbf{f}_2$. In fact,

$$D_{00}(R_2^{-1} R_1) = P_\nu(\cos \theta) \tag{10.25b}$$

and finally

$$W = (2j + 1)^2 \sum_{L_1 L_1'} \sum_{L_2 L_2'} \sum_\nu (-)^{L_2' - L_1'} (j \parallel T_{L_1} \parallel j_1)(j \parallel T_{L_1'} \parallel j_1)^*$$

$$\times (j \parallel T_{L_2} \parallel j_2)(j \parallel T_{L_2'} \parallel j_2)^* c_{\nu 0}(L_1 L_1') c_{\nu 0}(L_2 L_2')$$

$$\times W(jjL_1 L_1'; \nu j_1) \, W(jjL_2 L_2'; \nu j_2) \, P_\nu(\cos \theta) \tag{10.26}$$

The value of ν is limited by the triangular conditions $\Delta(jj\nu)$, $\Delta(L_1 L_1'\nu)$, $\Delta(L_2 L_2'\nu)$.

So far nothing has been said about the nature of the radiations, and (10.26) is quite general. From (7.38) it follows that for gamma rays

$$A_{L,-M} = (2L + 1)^{\frac{1}{2}} e^{i\varphi}$$

where $e^{i\varphi}$ is a phase factor which depends on whether the radiation is electric or magnetic in character. For either type of radiation

$$c_{\nu 0}(LL') = \sum_{M = \pm 1} (-)^{L-M}[(2L + 1)(2L' + 1)]^{\frac{1}{2}} C(LL'\nu; M, -M)$$

$$= (-)^{L-1}[(2L + 1)(2L' + 1)]^{\frac{1}{2}} C(LL'\nu; 1, -1)[1 + (-)^{L+L'-\nu}]$$

When pure electric or magnetic radiation is considered, the interfering multipoles must have the same parity so that $|L - L'|$ is an even number. Hence ν is even, and the correlation function is symmetrical about $\theta = \pi/2$. This result is not restricted to pure radiations because for the interference terms we must examine the phase factor $e^{i\varphi}$. Then, since $e^{i\varphi} \sim p = \pm 1$, it follows that the square bracket in $c_{\nu 0}$ is replaced by

$$1 - (-)^{L+L'-\nu}$$

and hence, since $|L - L'|$ is now an odd integer, ν is still even. This is, in fact, a general property of all correlations except the rather academic case in which the circular polarization of two gamma rays is measured. The even character of ν should be borne in mind in connection with the triangular conditions limiting ν_{max}. For $j = \frac{1}{2}$ we obtain $\nu_{max} = 0$ which implies isotropy. This confirms an earlier statement.

For simplicity we restrict our attention to pure radiations. Then for electric or magnetic radiations the correlation function is

$$W = \sum_\nu A_\nu P_\nu(\cos \theta) \tag{10.27}$$

$$A_\nu = F_\nu(L_1 j_1 j) \, F_\nu(L_2 j_2 j) \tag{10.27a}$$

$$F_\nu(Lj_1j) = (-)^{j_1-j-1}(2j+1)^{\frac{1}{2}}(2L+1)\ C(LL\nu;1,-1)\ W(jjLL;\nu j_1)$$

$$(10.27b)$$

Here ν_{max} is the smallest of $2j$, $2L_1$, $2L_2$.

For spinless particles $M = 0$ only and we need to replace $C(LL\nu;1,-1)$ by $-C(LL\nu;00)$. It is useful to recall the relation (3.30) in this connection. For interference terms $|L - L'|$ is always an even integer. Here ν_{max} can be zero if either L_1 or $L_2 = 0$. That this should give isotropy is physically obvious.

For mixed transitions involving gamma rays, we define

$$f_\nu(LL'j_1j) = (-)^{j_1-j-1}[(2j+1)(2L+1)(2L'+1)]^{\frac{1}{2}}$$

$$\times C(LL'\nu;1,-1)\ W(jjLL';\nu j_1) \quad (10.27c)$$

in place of $F_\nu(Lj_1j)$. Then with

$$A_\nu = f_\nu(L_1L_1'j_1j)\ f_\nu(L_2L_2'j_2j)$$

the mixed correlation function is

$$W = \sum_{L_1L_1'} \sum_{L_2L_2'} \sum_\nu a(L_1L_1';L_2L_2')\ A_\nu P_\nu$$

The constants $a(L_1L_1';L_2L_2')$, which are simply the product of four reduced matrix elements, are obviously symmetrical with respect to interchange of primed and unprimed letters, as well as subscripts 1 and 2. In the terms with $L_1 \neq L_1'$ or $L_2 \neq L_2'$, the smallest value of ν is the smallest even integer that is equal to or larger than $|L_1 - L_1'|$ or $|L_2 - L_2'|$. Since this excludes $\nu = 0$, these interference terms do not contribute to the total intensity. Averaging over all directions we find

$$W_{avg} = \sum_{L_1L_2} a(L_1L_1;L_2L_2)$$

so that the squares of the reduced matrix elements represent the relative intensities of the respective radiations. These remarks obviously apply to the emission of alpha particles or other types of radiation.

34. EMISSION OF ALPHA PARTICLES BY AN ORIENTED NUCLEUS

With the framework of the angular correlation theory as presented in section 33 before us, it is a comparatively simple matter to discuss the angular distribution of radiation emitted by nuclei oriented by external fields. This will always imply a low temperature since the thermal energy kT must be of order of the splittings produced by the coupling of the nuclear spin to the external field. Generally this means $T \lesssim 1°$ K. In this section we consider aligned nuclei so that the external field is either the crystalline field producing anisotropic hyperfine coupling

(Bleaney alignment [8]) or the inhomogeneous electric field coupled to the nuclear quadrupole moment (Pound alignment [9]). The discussion will deal with emission of alpha particles. For emission of gamma rays only a trivial change is needed, as the discussion of the preceding section indicates.

As the discussion of section 32 indicates, the angular distribution of particles emitted from the nuclear (or any other) source will in general be anisotropic if the population of the substates of the emitting level is not uniform. The anisotropy is defined relative to the direction or orientation of the field (or crystalline axes associated with the field) which is responsible for a splitting of the levels. If, in the presence of a perturbation, the energy $E(m_i)$ of the emitting state is dependent on the magnetic quantum number m_i, then the population of this substate is, at temperature T,

$$p(m_i) = e^{-E(m_i)/kT} \tag{10.28}$$

If we consider the various moments

$$\langle m_i^n \rangle = \frac{\sum_{m_i} m_i^n \, e^{-E(m_i)/kT}}{\sum_{m_i} e^{-E(m_i)/kT}} \tag{10.29}$$

it is clear that the effect discussed in section 32 depends only on the first moment $\langle m_i \rangle$. Thus, if a nucleus is polarized and has a net magnetic moment, its cross section for absorption of polarized (s) neutrons is changed as we have seen. But, if only $\langle m_i \rangle$ or if only average values of odd powers of m_i are changed from the non-oriented value, radiation emitted by the nucleus will be isotropic, as the following development will bring out explicitly.

For the transition $j_i, m_i \rightarrow j_f, m_f$ with the emission of an alpha particle with the quantum number L the transition probability is clearly proportional to

$$W \sim \sum_{m_i m_f} p(m_i) \, | \, (j_f m_f | H_\alpha(LM) | j_i m_i) \, |^2 \tag{10.30}$$

and we recognize that $m_i = m_f + M$, so that the M sum may be regarded as already accomplished. Now

$$H_\alpha(LM) = \sum_M (-)^M \, Y_{L,-M}(\mathbf{f}) \, T_{LM}(\mathbf{X}) \tag{10.31}$$

where $\mathbf{X}$ contains nuclear coordinates and $\mathbf{f}$ is the alpha-particle unit propagation vector. As was already mentioned, the sum over M can be

[8] B. Bleaney, *Proc. Phys. Soc. (London)* **A 64,** 315 (1951); *Phil. Mag.* **42,** 441 (1951).
[9] R. V. Pound, *Phys. Rev.* **76,** 1410 (1949).

dropped. In this connection it is important to notice that the m_i, m_f sums are incoherent. As a consequence

$$W \sim \sum_{m_i m_f} p(m_i) \, | \, (j_f m_f \, | \, T_{LM} \, | j_i m_i) \, |^2 \, | \, Y_{L,-M} \, |^2$$

We use the Wigner–Eckart theorem, that is,

$$(j_f m_f \, | \, T_{LM} \, | j_i m_i) = C(j_i L j_f; m_i, m_f - m_i)(j_f \, \| \, T_L \, \| \, j_i)$$

and, for simplicity, consider only one L-value. Hence the reduced matrix element is of no further interest because it is a scale factor only. For the more general case where several L-values interfere, the reduced matrix elements are easily restored.

We also use (4.31) and (4.32) to write

$$| \, Y_{L,-M} \, |^2 = | \, Y_{LM} \, |^2$$
$$= (-)^M \sum_\nu \frac{2L+1}{[4\pi(2\nu+1)]^{\frac{1}{2}}} C(LL\nu; 00) \, C(LL\nu; M, -M) \, Y_{\nu 0}(\mathfrak{k})$$

Then, since $Y_{\nu 0} = \left(\dfrac{2\nu+1}{4\pi}\right)^{\frac{1}{2}} P_\nu$, we obtain the result

$$W \sim \sum_\nu C(LL\nu; 00) \, P_\nu \sum_{m_i m_f} (-)^M \, C(LL\nu; M, -M) \, p(m_i)$$
$$\times \, [C(j_i L j_f; m_i M)]^2 \, \delta_{M, m_f - m_i}$$

This shows that ν must be an even integer; see (3.22). Again, for (unpolarized) gamma-ray emission we replace $C(LL\nu; 00)$ by $C(LL\nu; 1, -1)$. This also applies to the result (10.32) below.

Since $p(m_i)$ is not a particular simple function of m_i, it is advantageous to carry out the sum over m_f first, keeping m_i fixed. This sum is

$$S = \sum_{m_f} (-)^{m_f - m_i} \, C(LL\nu; M, -M) \, [C(j_i L j_f; m_i M)]^2 \, \delta_{M, m_f - m_i}$$

In this sum we use

$$C(j_i L j_f; m_i M) = (-)^{j_i - m_i} \left(\frac{2j_f+1}{2L+1}\right)^{\frac{1}{2}} C(j_i j_f L; m_i, -m_f)$$

and

$$C(LL\nu; M, -M) = C(LL\nu; -M, M)$$

since $(-)^{2L-\nu} = 1$. Consequently,

$$C(j_i j_f L; m_i, -m_f) \, C(LL\nu; -M, M) = \sum_t [(2t+1)(2L+1)]^{\frac{1}{2}}$$
$$\times \, C(j_f L t; -m_f M) \, C(j_i t\nu; m_i, -m_i) \, W(j_i j_f \nu L; Lt)$$

by (6.4b). Then the sum S reduces to

$$S = \sum_t [(2t + 1)(2j_f + 1)]^{\frac{1}{2}} C(j_i t \nu; m_i, -m_i) W(j_i j_f \nu L; Lt)$$

$$\times \sum_{m_f} C(j_f Lt; -m_f, m_f - m_i) C(j_i L j_f; m_i, m_f - m_i)(-)^{j_f - m}$$

In the second C-coefficient of the m_f sum we write

$$C(j_i L j_f; m_i, m_f - m_i) = \left(\frac{2j_f + 1}{2j_i + 1}\right)^{\frac{1}{2}} (-)^{L + M_f - m_i} C(j_f L j_i; -m_f, m_f - m_i)$$

by (3.17a). Then the m_f sum is

$$\sum_{m_f} \cdots = \left(\frac{2j_f + 1}{2j_i + 1}\right)^{\frac{1}{2}} (-)^{L + j_f - m_i} \delta_{t j_i}$$

and

$$S = (2j_f + 1)(-)^{L + j_f - m_i} C(j_i j_i \nu; m_i, -m_i) W(j_i j_f \nu L; L j_i)$$

$$= (-)^{j_i - m_i}(2j_f + 1) C(j_i j_i \nu; m_i, -m_i) W(j_i j_i LL; \nu j_f)$$

by (6.10b).

Now the angular distribution of the alpha particles becomes

$$W(\theta) \sim \sum G_\nu C(LL\nu; 00) W(j_i j_i LL; \nu j_f) P_\nu \qquad (10.32)$$

where the properties of the emitting state are contained in the parameters

$$G_\nu = \sum_{m_i} (-)^{j_i - m_i} p(m_i) C(j_i j_i \nu; m_i, -m_i) \qquad (10.33)$$

These parameters G_ν are special cases of statistical tensors introduced by Fano.[10] In (10.33) we have again dropped multiplicative constants. The factor $(-)^{j_i}$ is kept so that one is assured that the phase is real: $j_i - m_i$ is an integer. Of course, $0 \leq \nu \leq \nu_{max}$ and ν_{max} is the smaller of $2j_i$ or $2L$. If j_i is a half-integer, the pertinent upper limit is $2j_i - 1$ or $2L$.

As a check we may first see that for $p(m_i) = $ a constant, $G_\nu = 0$ unless $\nu = 0$. Of course, this would then mean isotropy. In this case G_ν is proportional to

$$G_\nu^{(0)} = \sum_{m_i} (-)^{j_i - m_i} C(j_i j_i \nu; m_i, -m_i)$$

But

$$C(j_i j_i 0; m_i, -m_i) = \frac{(-)^{j_i - m_i}}{(2j_i + 1)^{\frac{1}{2}}}$$

[10] U. Fano, *Natl. Bur. Standards Rept. 1214* (1951).

and therefore

$$G_\nu^{(0)} = (2j_i + 1)^{\frac{1}{2}} \sum_{m_i} C(j_i j_i \nu; m_i, -m_i) \, C(j_i j_i 0; m_i, -m_i)$$

$$= \delta_{\nu 0}(2j_i + 1)^{\frac{1}{2}} \tag{10.34}$$

which is the desired result.

It is also clear that $j_i > \frac{1}{2}$ is necessary for anisotropy. This follows from the condition $\Delta(j_i j_i \nu)$ which means $0 \leq \nu \leq 2j_i$. For $j_i = \frac{1}{2}$, $\nu_{max} = 1$, and the even character of ν forces the conclusion $\nu = 0$.

Now consider the polarization produced by capturing slow, polarized neutrons. In this case $p(m_i)$ is proportional to $[C(j\frac{1}{2}j_i; mm_s)]^2$ where j is the spin of the target nucleus and m its projection quantum number. Thus $p(m_i)$ is a linear function of $m_i = m + m_s$

$$p(m_i) = a + bm_i \tag{10.35}$$

The term independent of m_i contributes nothing to the anisotropy as we have just seen. Then we need only examine

$$G_\nu^{(1)} = \sum_\nu (-)^{j_i - m_i} m_i \, C(j_i j_i; m_i, -m_i)$$

But we know that

$$C(j_i 1 j_i; m_i 0) = m_i [j_i(j_i + 1)]^{-\frac{1}{2}}$$

$$= (-)^{j_i - m_i} [3/(2j_i + 1)]^{\frac{1}{2}} \, C(j_i j_i 1; m_i, -m_i)$$

$$G_\nu^{(1)} = \left[\frac{3j_i(j_i + 1)}{2j_i + 1} \right]^{\frac{1}{2}} \sum_{m_i} C(j_i j_i \nu; m_i - m_i) \, C(j_i j_i 1; m_i, -m_i)$$

$$= \left[\frac{3j_i(j_i + 1)}{2j_i + 1} \right]^{\frac{1}{2}} \delta_{\nu 1} \tag{10.36}$$

But $\nu = $ even only. Hence this term vanishes, and we have isotropy again. The capture of polarized s-neutrons will, however, lead to the emission of circularly polarized gamma rays.[11]

We can easily generalize this result. Suppose

$$p(m_i) = p_0 + q(m_i) \tag{10.37}$$

where p_0 is independent of m_i and $q(m_i)$ is odd:

$$q(m_i) = -q(-m_i) \tag{10.37'}$$

Then we can disregard p_0 as contributing only an isotropic background term. The remaining term is

$$G_\nu(q) = \sum_{m_i} (-)^{j_i - m_i} q(m_i) \, C(j_i j_i \nu; m_i, -m_i)$$

[11] L. C. Biedenharn, M. E. Rose, and G. B. Arfken, *Phys. Rev.* **83**, 683 (1951).

But, since $-j_i \leq m_i \leq j_i$, we can change the sign of the summation variable m_i. Thus,

$$G_\nu(q) = - \sum_{m_i} (-)^{j_i + m_i} q(m_i) \, C(j_i j_i \nu; -m_i m_i)$$

$$= (-)^{2j_i - \nu + 1} \sum_{m_i} (-)^{j_i + m_i} q(m_i) \, C(j_i j_i \nu; m_i, -m_i)$$

by $(3.16a)$. Since $(-)^\nu = 1$, this gives the result

$$G_\nu(q) = -G_\nu(q) = 0 \tag{10.38}$$

Of course, the constant term p_0 in (10.37) means that G_ν vanishes except for $\nu = 0$, as in (10.34). Therefore, the population distribution (10.34) implies an isotropic angular distribution. Consequently, for anisotropy it is necessary that $p(m_i)$ have a part which is even in m_i and is dependent on m_i. This confirms the statement made earlier that it is not sufficient to induce a magnetic moment in the source. An orientation in which $p(m_i)$ is even, i.e., one that is invariant to sign reversal of j_{zi}, does give anisotropy. This is the case of alignment, and we can produce this by quadrupole coupling of the nuclear quadrupole moment to the inhomogeneous electric field arising from surrounding electrons (Pound alignment [9]) or by anisotropic hyperfine coupling (Bleaney alignment [8]).

As soon as we recognize that the interaction which produces the alignment may depend on the spin operators of both nucleus and surrounding electrons, it becomes necessary to revise the treatment of the angular distribution problem given above. Even if the dipole hyperfine coupling is of the form $\mathbf{J} \cdot \mathbf{S}$ and is isotropic and the alignment arises only from quadrupole coupling, it is necessary to recognize that, because of the coupling of the two spins, the projection quantum number of the nucleus is not a constant of the motion, and we must treat the system of nucleus plus electrons as a whole. Thus, the foregoing discussion, in which the nucleus is regarded as an isolated system, is not quite complete. It will turn out that the only revision that is necessary is a more or less obvious generalization of the meaning of the statistical tensors G_ν.

Usually, at the low temperatures envisaged, the electric field (Stark) splittings are very large, and the ground state of the ion is an orbital singlet.[12] That is, $L = 0$ effectively for the electrons. Thus, the angular momentum for the electron is represented by S. However, it is not necessarily true that $\mathbf{F} = \mathbf{J} + \mathbf{S}$ is conserved.[12] Instead the ion is represented by a pair of quantum numbers, which we call a and b, and the wave functions which diagonalize the total energy, as well as the opera-

[12] B. Bleaney, *Physica* **XVII**, 175 (1951). Conservation of $\mathbf{F}$ means that $\mathbf{F}^2$ is diagonal.

tors corresponding to a and b, are Ψ_{ab}. These are obtainable from the uncoupled representation $\psi_{jm_j}\,\psi_{sm_s}$ by a unitary transformation

$$\Psi_{ab} = \sum_{m_j m_s} U_{ab,m_j m_s}\,\psi_{jm_j}\,\psi_{sm_s} \tag{10.39}$$

and the elements of the unitary matrix U are not necessarily C-coefficients. This is obvious, of course, if a magnetic field is present, but, as previously remarked, it is not necessary to invoke a magnetic field. Of course, a, b are eigenvalues of operators which commute with the Hamiltonian of the ion (in the ground state). We do not exclude the possibility that one of these may be the energy itself.

For the angular distribution we consider

$$W = \left\{ \sum_{ab}\sum_{a'b'} |\,(a'b'\,|\,H_\alpha\,|\,ab)\,|^2 \right\}_{\text{ave}} \tag{10.40}$$

where primes mean final state and ave means statistical average over the Boltzmann distribution for the initial state. Now, by using (10.39), we can write

$$W = \left\{ \sum_{ab}\sum_{a'b'} \left| \sum_{m_j m_s}\sum_{m'_j m'_s} U_{ab,m_j m_s}\,U^*_{a'b',m'_j m'_s}(j'm'_j s'm'_s\,|\,H_\alpha\,|\,jm_j sm_s) \right|^2 \right\}_{\text{ave}} \tag{10.41}$$

Since H_α is diagonal in the electron quantum numbers, we have that $s = s'$, $m_s = m'_s$, and we abbreviate

$$(j'm'_j s'm'_s\,|\,H_\alpha\,|\,jm_j sm_s) \equiv (j'm'_j\,|\,H_\alpha\,|\,jm_j)$$

The angular distribution function now becomes

$$W = \left\{ \sum_{ab}\sum_{a'b'} \sum_{m_s \bar{m}_s}\sum_{m_j \bar{m}_j}\sum_{m'_j \bar{m}'_j} U_{ab,m_j m_s}\,U^*_{a'b',m'_j m_s}\,U_{ab,\bar{m}_j \bar{m}_s}\,U_{a'b',\bar{m}'_j \bar{m}_s} \right.$$

$$\left. \times \; (j'm'_j\,|\,H_\alpha\,|\,jm_j)(j'm'_j\,|\,H_\alpha\,|\,jm_j)^* \right\}_{\text{ave}} \tag{10.42}$$

The a,b sum is readily carried out, using the unitary property of the U-matrix:

$$\sum_{ab} (U^{-1}_{m_j m_s,ab}\,U_{ab,\bar{m}_j \bar{m}_s})^* = \delta_{m_j \bar{m}_j}\,\delta_{m_s \bar{m}_s}$$

Also the a',b' sum is

$$\sum_{a'b'} U^{-1}_{m'_j m_s,a'b'}\,U_{a'b'\,\bar{m}'_j \bar{m}_s} = \delta_{m'_j \bar{m}'_j}\,\delta_{m_s \bar{m}_s}$$

Hence all magnetic quantum number sums are diagonal (no cross-terms), and we obtain

$$W = \left\{ \sum_{m_s} \sum_{m_j} \sum_{m_j'} |(j'm_j'|H_\alpha|jm_j)|^2 \right\}_{ave}$$

$$= \sum_{m_s} \sum_{m_j} \sum_{m_j'} p(m_s, m_j) |(j'm_j'|H_\alpha|jm_j)|^2 \qquad (10.43)$$

This differs from (10.30) only in the recognition that the population factors may depend on m_s as well as on m_j, and a sum over the electron projection quantum numbers m_s is involved as well. The subsequent development is exactly the same as that already employed, and the final result for pure angular momentum alpha particles is (10.32), but now the statistical tensors are defined by

$$G_\nu = \sum_{m_s} \sum_{m_i} (-)^{j_i - m_i} p(m_s, m_i) C(j_i j_i \nu; m_i, -m_i) \qquad (10.44)$$

and we have reverted to the previous notation.

Now if we recognize that

$$(-)^{j_i - m_i} C(j_i j_i \nu; m_i, -m_i) = \left(\frac{2\nu + 1}{2j_i + 1}\right)^{\frac{1}{2}} C(j_i \nu j_i; m_i 0)$$

it is seen that

$$G_\nu \sim \sum_{m_s m_i} p(m_s, m_i) (j_i m_i | T_{\nu 0} | j_i m_i) \qquad (10.45)$$

where $T_{\nu 0}$ is an element of an irreducible tensor of rank ν and is expressed entirely in the nuclear space. The matrix element in (10.45) is therefore expressible in the form $(jm_i s m_s | T_{\nu 0} | jm_i s m_s)$. If, for convenience, we normalize $p(m_s, m_i)$ so that

$$\sum_{m_s} \sum_{m_i} p(m_s, m_i) = 1$$

it will be recognized that

$$G_\nu = \frac{\text{Tr } T_{\nu 0} \, e^{-H/kT}}{\text{Tr } e^{-H/kT}} \qquad (10.46)$$

and the trace, designated by Tr, which is independent of the representation, can be calculated in the decoupled (or uncoupled) representation. This is the form of (10.45) in fact. In that case $\text{Tr} = \text{Tr}_E \text{Tr}_N$ where Tr_E is the summation over electronic, Tr_N is the summation over nuclear quantum numbers.

It remains to identify the operators $T_{\nu 0}$ which appear in (10.46). For $\nu = 0$ we find immediately

$$T_{00} = (2j_i + 1)^{-\frac{1}{2}} \qquad (10.47a)$$

For $\nu = 2$, we obtain from Table 4^3 of reference 13,

$$T_{20} = \left[\frac{45}{(2j-1)j(2j+1)(j+1)(2j+3)}\right]^{\frac{1}{2}}(J_z^2 - \tfrac{1}{3}\mathbf{J}^2)$$

$$= \left[\frac{180(2j-2)!}{(2j+3)!}\right]^{\frac{1}{2}}(J_z^2 - \tfrac{1}{3}\mathbf{J}^2) \tag{10.47b}$$

For $\nu = 4$ application of (3.18) gives

$$T_{40} = 210\left[\frac{(2j-4)!}{(2j+5)!}\right]^{\frac{1}{2}}[J_z^4 - \tfrac{1}{15}(3\mathbf{J}^4 - \mathbf{J}^2) - \tfrac{1}{7}(6\mathbf{J}^2 - 5)(J_z^2 - \tfrac{1}{3}\mathbf{J}^2)] \tag{10.47c}$$

Of course $\mathbf{J}^2$ can be replaced by $j(j+1)$ and $\mathbf{J}^4$ by $j^2(j+1)^2$. Using the expressions for

$$S_k = \sum_{-j}^{j} m^{2k} \tag{10.48}$$

which are easily evaluated using the Euler–Maclaurin sum formula [14] to give

$$S_0 = 2j + 1 \tag{10.48a}$$

$$S_1 = \tfrac{1}{3}(2j+1)j(j+1) \tag{10.48b}$$

$$S_2 = \tfrac{1}{15}j(j+1)(2j+1)(3\eta - 1) \tag{10.48c}$$

$$S_3 = \tfrac{1}{21}j(j+1)(2j+1)(3\eta^2 - 3\eta + 1) \tag{10.48d}$$

where $\eta = j(j+1)$, it is seen that in each case except for $\nu = 0$,

$$\text{Tr } T_{\nu 0} = 0$$

This is to be expected in general since

$$\sum_m (jm | T_{\nu 0} | jm) = (j \| T_\nu \| j) \sum_m C(j\nu j; m0)$$

$$= (2j+1)(2\nu+1)^{-\frac{1}{2}}(j \| T \| j)\,\delta_{\nu 0} \tag{10.49}$$

This shows that every irreducible tensor of non-zero rank has vanishing trace.[15]

[13] E. U. Condon and G. H. Shortley, *Theory of Atomic Spectra*, Cambridge University Press, 1935.

[14] For example, *Fundamental Formulas of Physics*, p. 70, Prentice-Hall, New York, 1955.

[15] This result can be used to show that, if the spin Hamiltonian is a sum of contractions of non-zero-rank tensors (in the spaces of spin operators $\mathbf{J}$ and $\mathbf{S}$), then the splittings produced do not shift the center of gravity of the levels. It also verifies the expected result that for very high temperatures the angular distributions become isotropic.

The evaluation of the traces in (10.46) for most spin Hamiltonians H is quite straightforward, using the techniques described in reference 1. In this connection the results given in (10.48a–d) are very useful. This particular phase of the problem will not be pursued further.

To extend these results to mixtures of different orbital angular momenta of the alpha particles we use the fact that

$$C(LL'\nu; 00) = C(L'L\nu; 00), \qquad L + L' + \nu \text{ even}$$

and

$$W(j_i j_i LL'; \nu j_f) = W(j_i j_i L'L; \nu j_f)$$

by (6.10a). Hence, if we define the symmetrized distribution function

$$W_{LL'} = \sum_{\nu} G_{\nu}\, C(LL'\nu; 00)\, W(j_i j_i LL'; \nu j_f) P_{\nu} = W_{L'L} \qquad (10.50)$$

then the angular distribution is

$$W(\theta) = \sum_{LL'} a_{LL'}\, W_{LL'} \qquad (10.51)$$

with adjustable constants $a_{LL'} = a_{L'L}$ and $a_{LL} \geq 0$. The distribution function (10.32) in which only one partial wave enters is a special case: $W(\theta) = W_{LL}$.

The meaning of the constants a_{LL} is seen at once by integrating over all angles. This gives

$$\int W(\theta)\, d\Omega = \frac{4\pi(-)^{j_i - j_f}}{2j_i + 1} \sum_{L} \frac{a_{LL}}{2L + 1} \qquad (10.52)$$

The normalization can be adjusted so that the factor multiplying the sum in (10.52) is positive. Then $a_{LL} \geq 0$ and $a_{LL}/(2L + 1)$ are the relative intensities of the Lth partial wave. For all L and L' we can show that

$$a_{LL'} = \pm(a_{LL} a_{L'L'})^{\frac{1}{2}} \cos(\sigma_L - \sigma_{L'})$$

except that for $L = L'$ only the upper sign can apply. Here σ_L is the total phase shift for the Lth partial wave

$$\sigma_L = \tfrac{1}{2}L\pi + \arg \Gamma\left[L + 1 + \frac{2(Z - 2)e^2}{\hbar v}\right] \qquad (10.53)$$

The first term in (10.53) is due to the centrifugal barrier. The second term arises from the Coulomb field; Z is the atomic number of the emitter, and v is the velocity of the alpha particle. It will be recognized that the parameters a_{LL} are analogous to the parameters $a(L_1 L_1; L_2 L_2)$ introduced at the end of section 33 in connection with the angular correlation of gamma rays. The latter are associated with the product of intensities of the two pure multipoles with angular momenta L_1 and L_2. The alpha-particle coefficients are defined differently so that $(2L + 1)^{-1}$ appears explicitly in (10.52). For the interference terms $(L \neq L')$ the major distinction between the two cases arises from the Coulomb phase shift which, of course, applies only to charged particles.

XI. ANGULAR DISTRIBUTIONS IN NUCLEAR REACTIONS

The problem of angular distributions in nuclear reactions is a more or less straightforward application of the principles discussed in section 33 in connection with angular correlation.[1] In this description of nuclear reactions there usually remains some parameters (elements of the scattering matrix) which play the role of weight factors for the various possible channel spins. In this chapter we place emphasis on particular coupling models which, in effect, would evaluate these constants—to within an overall multiplicative factor. For this purpose we first study the special coupling schemes in the following section.

35. j–j AND L–S COUPLING

In both atomic and nuclear structure it is important to know how the interaction energy between particles depends on the orientation of their angular momenta. In the literature two different coupling schemes have been used extensively. These are j–j and L–S coupling. They represent extreme cases, and the actual situation may require a deviation from the description provided by either picture.

Consider two particles with intrinsic spin s_1 and s_2 and orbital angular momenta l_1 and l_2. Then for this system j–j coupling is described by the coupling equations

$$\mathbf{j}_1 = \mathbf{l}_1 + \mathbf{s}_1$$
$$\mathbf{j}_2 = \mathbf{l}_2 + \mathbf{s}_2$$
$$\mathbf{J} = \mathbf{j}_1 + \mathbf{j}_2 \tag{11.1}$$

This means that there is a strong (spin-orbit) coupling between l_1 and s_1 as well as between l_2 and s_2. Beside $\mathbf{s}_1^2$, $\mathbf{s}_2^2$, l_1^2, and l_2^2 the constants of the motion are

$$\mathbf{J}^2, J_z, \mathbf{j}_1^2, \mathbf{j}_2^2$$

where $J(J+1)$, $j_1(j_1+1)$, $j_2(j_2+1)$, and M are the eigenvalues of $\mathbf{J}^2$, $\mathbf{j}_1^2$, $\mathbf{j}_2^2$, and J_z, respectively.

---

[1] W. T. Sharp, J. M. Kennedy, B. J. Sears, and M. G. Hoyle, *Chalk River Rept.* CRT-556.

The case of L–S coupling is characterized by a strong coupling between l_1 and l_2 and also between s_1 and s_2. An example of the l_1–l_2 coupling is provided by any interaction energy $V(r_{12})$ depending on the interparticle distance r_{12}. The energy V can be expanded into Legendre polynomials $P_l(\cos \theta_{12})$, which has the form of a scalar product of two tensors:

$$P_l(\cos \theta_{12}) = \frac{4\pi}{2l + 1} \sum_m Y_{lm}(\theta_1 \varphi_1) \, Y_{lm}^*(\theta_2 \varphi_2)$$

and by (6.21) this is diagonal in the quantum number corresponding to the resultant $\mathbf{L} = l_1 + l_2$. In any event the L–S coupling equations are

$$\mathbf{L} = l_1 + l_2$$

$$\mathbf{S} = \mathbf{s}_1 + \mathbf{s}_2$$

$$\mathbf{J} = \mathbf{L} + \mathbf{S} \tag{11.2}$$

The constants of the motion are l_1^2, l_2^2, s_1^2, s_2^2, and

$$\mathbf{J}^2, J_z, \mathbf{L}^2, \mathbf{S}^2$$

where $L(L + 1)$ and $S(S + 1)$ are eigenvalues of $\mathbf{L}^2$ and $\mathbf{S}^2$.

In explicit form the wave functions are:

j–j Coupling

$$\Psi_{JM} = \sum_{m_1} C(j_1 j_2 j; m_1, M - m_1) \, \psi_{j_1 m_1} \psi_{j_2, M - m_1}$$

and

$$\psi_{j_1 m_1} = \sum_{\tau_1} C(l_1 s_1 j_1; m_1 - \tau_1, \tau_1) \, \psi_{l_1, m_1 - \tau_1} \psi_{s_1 \tau_1} \tag{11.3}$$

and an exactly similar equation for $\psi_{j_2, M - m_1}$.

L–S Coupling

$$\Phi_{JM} = \sum_m C(LSJ; m, M - m) \, \Psi_{Lm} \Psi_{S, M - m}$$

$$\Psi_{Lm} = \sum_\mu C(l_1 l_2 L; \mu, m - \mu) \, \psi_{l_1 \mu} \psi_{l_2, m - \mu}$$

$$\Psi_{S, M - m} = \sum_\tau C(s_1 s_2 S; \tau, M - m - \tau) \, \psi_{s_1 \tau} \psi_{s_2, M - m - \tau} \tag{11.4}$$

Since in these two types of coupling we are combining four angular momenta l_1, l_2, s_1, and s_2 to form a resultant $\mathbf{J}$, it is clear that we are confronted with a somewhat more complicated situation than that considered in section 22, where the coupling of three angular momenta to form a given resultant $\mathbf{J}$ was treated. Just as in that case, we are now concerned with the connection between two representations Ψ and Φ

which represent alternative ways of combining the original four angular momenta.

We write

$$\Phi_{JM}(LS) = \sum_{j_1 j_2} T(j_1 j_2; LS) \, \Psi_{JM}(j_1 j_2) \tag{11.5}$$

The matrix elements $T(j_1 j_2; LS)$ then effect a recoupling from the j–j to the L–S schemes.

In (11.5) we have added the arguments L, S and j_1, j_2 to the wave functions to distinguish the two schemes. The coefficients $T(j_1 j_2; LS)$ obviously form the elements of a unitary matrix. The only other information that is obvious is that $T(j_1 j_2; LS)$ must reduce to an R matrix (see section 22) when one of l_1, l_2, s_1, s_2 vanishes. In practice $s_1 = s_2 = \frac{1}{2}$ in cases of interest. However, if $l_1 = 0$ for example, $j_1 = s_1$ and $L = l_2$, and only three angular momenta are being coupled. Also, if one of L, S, or J vanishes, it follows that there are only two independent coupling equations in (11.1) or (11.2), so that T should again reduce to R.

To calculate the T-matrix we form the scalar product $(\Psi, \Phi) = (\Phi, \Psi)$; that is,

$$T(j_1 j_2; LS) = (\Psi, \Phi) = \sum C(j_1 j_2 j; m_1, M - m_1) \, C(l_1 s_1 j_1; m_1 - \tau_1, \tau_1)$$

$$\times \, C(l_2 s_2 j_2; M - m_1 - \tau_2, \tau_2) \, C(LSJ; m, M - m) \, C(l_1 l_2 L; \mu, m - \mu)$$

$$\times \, C(s_1 s_2 S; \tau, M - m - \tau) \, \delta_{\tau \tau_1} \, \delta_{M - m - \tau, \tau_2} \, \delta_{\mu, m_1 - \tau_1} \, \delta_{m - \mu, M - m_1 - \tau_2} \tag{11.6}$$

The sum in (11.6) is over m_1, τ_1, τ_2, m, μ, and τ. The δ-symbols arising from the scalar products of orthonormal wave functions permit the τ_1, τ_2, and m_1 sums to be done immediately. That is, we set

$$\tau_1 = \tau$$

$$\tau_2 = M - m - \tau$$

$$m_1 = \mu + \tau_1 = \mu + \tau$$

Then we have a triple sum of the product of six C-coefficients.

$$T(j_1 j_2; LS) = \sum_{\tau m \mu} C(j_1 j_2 j; \mu + \tau, M - \mu - \tau) \, C(L_1 S_1 j_1; \mu \tau)$$

$$\times \, C(l_2 s_2 j_2; m - \mu, M - m - \tau) \, C(LSJ; m, M - m)$$

$$\times \, C(l_1 l_2 L; \mu, m - \mu) \, C(s_1 s_2 S; \tau, M - m - \tau) \tag{11.7}$$

We number these C-coefficients (1) $\cdots$ (6) in reading order. Then the product of (1) and (2) is

$$(2)(1) = \sum_{t_1} [(2t_1 + 1)(2j_1 + 1)]^{\frac{1}{2}} \, C(s_1 j_2 t_1; \tau, M - \mu - \tau)$$

$$\times \, C(l_1 t_1 j; \mu, M - \mu) \, W(l_1 s_1 j j_2; j_1 t_1) \tag{11.8a}$$

by the usual recoupling rule (6.4b). Notice that one of these new C-coefficients is independent of τ. This means that τ occurs in three C-coefficients at this stage. We take two of these, (3) and (6), and eliminate τ from one of the resulting C-coefficients:

$$(3)(6) = \left(\frac{2j_2 + 1}{2s_2 + 1}\right)^{\frac{1}{2}}(-)^{l_2-m+\mu}\, C(l_2j_2s_2;\, m-\mu,\, -M+\mu+\tau)$$

$$\times\, C(s_2s_1S;\, -M+m+\tau,\, -\tau)$$

$$= \left(\frac{2j_2 + 1}{2s_2 + 1}\right)^{\frac{1}{2}}(-)^{l_2-m+\mu}\sum_{t_2}[(2s_2 + 1)(2t_2 + 1)]^{\frac{1}{2}}$$

$$\times\, C(j_2s_1t;\, \mu+\tau-M,\, -\tau)\, C(l_2t_2S;\, m-\mu,\, \mu-M)$$

$$\times\, W(l_2j_2Ss_1;\, s_2t_2) \tag{11.8b}$$

Here the first equality results from (3.16a, b, and c). We now combine the remaining two C-coefficients which contain τ. These are the first on the right sides of (11.8a) and (11.8b). Calling these (7) and (8), we get

$$\sum_{\tau}(7)(8) = \sum_{\tau} C(j_2s_1t_1;\, -M+\mu+\tau,\, -\tau)\, C(j_2s_1t_2;\, \mu+\tau-M,\, -\tau)$$

$$= \delta_{t_1t_2}$$

This permits the t_2-sum to be done. So far we have

$$T(j_1j_2;\, LS) = [(2j_1 + 1)(2j_2 + 1)]^{\frac{1}{2}}\sum_{t_1}(2t_1 + 1)\, W(l_1s_1jj_2;\, j_1t_1)$$

$$\times\, W(l_2j_2Ss_1;\, s_2t_1)\sum_{m\mu}(-)^{l_2-m+\mu}[(4)(5)(9)(10)]_{t_2=t_1} \tag{11.9}$$

where (9) and (10) denote the non-τ-dependent C-coefficients in (11.8a) and (11.8b), respectively. For the sum over μ we need consider only (5), (9), and (10). For the product of the first two we find

$$5)(9) = \left(\frac{2L + 1}{2l_1 + 1}\right)^{\frac{1}{2}}(-)^{l_2+m-\mu}\, C(Ll_2l_1;\, -m,\, m-\mu)\, C(l_1t_1j;\, \mu,\, M-\mu)$$

$$= \left(\frac{2L + 1}{2l_1 + 1}\right)^{\frac{1}{2}}(-)^{m-\mu-L+l_1}\, C(Ll_2l_1;\, m,\, \mu-m)\, C(l_1t_1j;\, \mu,\, M-\mu)$$

$$= (2L + 1)^{\frac{1}{2}}(-)^{m-L+l_1-\mu}\sum_{t_3}(2t_3 + 1)^{\frac{1}{2}}\, C(l_2t_1t_3;\, \mu-m,\, M-\mu)$$

$$\times\, C(Lt_3J;\, m,\, M-m)\, W(Ll_2Jt_1;\, l_1t_3)$$

The two C-coefficients appearing in the last line will be called (11) and

(12), respectively. Then, remembering that the phase factors combine so that μ drops out, we can write

$$\sum_\mu (10)(11) = \sum_\mu C(l_2 t_1 S; m-\mu, \mu-M) \, C(l_2 t_1 t_3; \mu-m, M-\mu)$$

$$= (-)^{l_2 + t_1 - S} \, \delta_{S t_3}$$

Thus, the t_3-sum is done. Those remaining are the m- and t_1-sums. We now have

$$T(j_1 j_2; LS) = [(2j_1 + 1)(2j_2 + 1)(2L + 1)(2S + 1)]^{\frac{1}{2}} X \quad (10.10)$$

where

$$X = (-)^{l_1 - L + l_2}(-)^{l_2 - S} \sum_{t_1} (2t_1 + 1)(-)^{t_1} \, W(l_1 s_1 J j_2; j_1 t_1)$$

$$\times W(l_2 j_2 S s_1; s_2 t_1) \, W(L l_2 J t_1; l_1 s) \sum_m C(LSJ; m, M-m)$$

$$\times C(LSJ; m, M-m)$$

Of course, the last sum gives unity. Dropping the subscript on t_1 and using the symmetry relations for the Racah coefficients, we find

$$X \equiv X(l_1 s_1 j_1; l_2 s_2 j_2; LSJ)$$

$$= (-)^\sigma \sum_t (2t + 1) \, W(s_1 l_2 j_1 L; tl_1) \, W(l_2 s_1 j_2 S; ts_2) \, W(Lj_1 Sj_2; tJ)$$

$$\qquad\qquad (11.11)$$

$$\sigma = l_1 + s_1 + j_1 + l_2 + s_2 + j_2 + L + S + J \qquad (11.11')$$

This quantity is generally referred to as the X-coefficient. It is important to remember that it is associated with the addition of four angular momenta to form a fifth, just as the simpler W-coefficient is associated with the addition of three angular momenta to form a fourth. It will also be seen that, if the nine arguments are arranged in a square array:

$$\begin{bmatrix} l_1 & s_1 & j_1 \\ l_2 & s_2 & j_2 \\ L & S & J \end{bmatrix}$$

then the entries in any column or any row form a triangle. Hence, it is not surprising that, within a phase, any pair of rows or columns can be interchanged. When this is done, X changes by the factor $(-)^\sigma$.

To find the form of the X-coefficient when one of the nine parameters vanishes we can choose any one of the arguments in (11.11) to be zero. The position of the vanishing element can be shifted by using the rule just given for interchanging rows and/or columns. If in (11.11) we set

$s_1 = 0$, the sum reduces to one term, $t = l_2$, and two of the Racah coefficients are evaluated by (6.12). The result can then be put in the form

$$X(abc;dec;gg0) = \frac{(-)^{c+g-a-e} W(abde;cg)}{[(2c+1)(2g+1)]^{\frac{1}{2}}} \tag{11.12}$$

This is in confirmation of remarks made above.

Although the X-coefficient appears here as a factor in the elements of the unitary matrix connecting two coupling schemes, it will be apparent that it will also appear in many other applications. For instance, the X-coefficient appears in the single-particle j–j matrix elements of the operator $\boldsymbol{\sigma} \cdot \mathbf{T}_{JLM}$. The wave functions in this case are given by (9.3). The operator in question occurs, for example, in the tensor type of coupling in beta decay. The precise nature of the X-coefficient is apparent from the triangular relations existing between the relevant parameters: $\Delta(J1L)$ which corresponds to the construction of the tensor of rank J from those of rank L and 1. The latter are the spherical harmonic of order L and the solid harmonic (argument σ) of order 1. $\Delta(j_1 s_1 l_1)$ and $\Delta(j_2 s_2 l_2)$ where the subscripts 1 and 2 refer to the two states defining the matrix element. Here $s_1 = s_2 = \frac{1}{2}$. The matrix element in configuration space is of the form (4.33') and gives $\Delta(Ll_1 l_2)$. The matrix element in spin space gives $\Delta(1\frac{1}{2}\frac{1}{2})$. Finally the Wigner–Eckart theorem shows that a relation $\Delta(Jj_1 j_2)$ must exist. From these remarks it is fairly evident that $X(J1L; j_1 s_1 l_1; j_2 s_2 l_2)$ will occur in the matrix element in question.

Finally, we repeat that the rows and columns of the matrix of which $T(j_1 j_2; LS)$ or X are the elements are labeled by $j_1 j_2$ and LS. That is, each row (column) has a double label. The orthonormality rules for unitary matrices can be written down immediately, and this is left as an exercise. Note that the X- and T-coefficients are real.

The X-coefficients, first studied by Wigner,[2] have been investigated by Fano[3] and by Schwinger.[4] Numerical tables have been prepared by Sharp et al.[1] and by Matsunobu and Takebe.[5] Their properties have been discussed by Sharp[6] and by Arima et al.[7]

[2] E. P. Wigner, On the Matrices Which Reduce the Kronecker Products of Representations of Simply Reducible Groups (unpublished); E. P. Wigner, *Am. J. Math.* **63**, 57 (1941).

[3] U. Fano, *Natl. Bur. Standards Rept. 1214* (1951).

[4] J. Schwinger, On Angular Momentum, *Nuclear Development Corp. of America Rept. NYO-3071* (January 26, 1952).

[5] H. Matsunobu and H. Takebe, *Prog. Theoret. Phys.* **14**, 589 (1955).

[6] W. T. Sharp, Some Formal Properties of the 12-j Symbol, *Chalk River Rept. TPI-81. Bull. Am. Phys. Soc. ser. II*, **1**, 210 (1956). It is clear that the recoupling coefficients for the addition of $n-1$ angular momenta to form an nth is characterized by $3(n-2)$ parameters.

[7] A. Arima, H. Horie and Y. Tanabe, *Prog. Theoret. Phys.* **11**, 143 (1954).

There are more complicated quantities associated with the addition of five and more angular momenta.[6] However, we shall not pursue this subject further.

36. ANGULAR MOMENTUM COUPLING IN NUCLEAR REACTIONS

As was previously emphasized, in a nuclear reaction wherein formation of a compound nuclear state of angular momentum J is involved, there are a number of angular momenta which are added to produce this resultant J, and the manner of their coupling is decisive for the angular distribution of outgoing particles. Suppose we consider the formation of a compound state when particles X bombard nucleus A to form the compound nucleus C. Then we have the following angular momenta to consider:

s_i: The intrinsic spin of the incident particle X.

l_i: The internal orbital angular momentum of X. For all cases of interest $l_i = 0$. For example, this is true for neutrons, protons, alpha particles, and, omitting small tensor force effects, it is true for H^2, H^3 and He^3. It would not be true for some heavier particles like Li^7.

l: The orbital angular momentum of the relative motion of X and A.

S_t: The intrinsic spin of A, the target nucleus.

L_t: The internal orbital angular momentum of A.

J_t: The total angular momentum of A.

J: The total angular momentum of C.

Clearly $\mathbf{J}$, as an operator,[8] is the vector sum of $\mathbf{s}_i$, $\mathbf{S}_t$, $\mathbf{L}_t$, and $\mathbf{l}$. Here, and henceforth, we put $l_i = 0$. But the order in which these vectors are to be added is not known beforehand. This would depend on what angular momenta are constants of the motion, that is, which momenta commute with the Hamiltonian of the problem. Since we do not know the nuclear forces operative here well enough to answer this question, the procedure is to try different extreme coupling assumptions and compare the results with experiment. This was first done by Christy.[9] Results of a somewhat more general nature appear in the work of Satchler [10] and of Sharp et al.,[1] although in these latter two references no attempt is made to compare the results of the calculations with experiment. Actually, it is not necessary that the observed results conform to any of the simple coupling schemes given below, and as stated some sort of intermediate coupling may be called for. The answer will surely be de-

[8] We shall use the same letters for operators and eigenvalues, and these two are distinguished by the fact that the former is printed in boldface type.

[9] R. Christy, *Phys. Rev.* **89**, 839 (1953).

[10] G. R. Satchler, *Proc. Phys. Soc.* **A 66**, 1081 (1953).

pendent on what part of the periodic table is under consideration. Even so, there is some value in presenting the following discussion because, as will be evident, the correct description can be represented in terms of a linear combination of possible results in any one of the coupling schemes. This implies the introduction of constants in the linear combination, and these constants are generally to be regarded as parameters to be adjusted so that a fit of the observations can be obtained. The situation is somewhat reminiscent of but not altogether like that encountered when a nucleus emits gamma rays which are a mixture of two multipoles. There, too, the experiments serve to determine (coherent) mixture constants. Here it is the coupling scheme that constitutes the question of interest.

To be specific there are three coupling schemes which can be discussed. One of these is the channel spin representation. Here one adds $j_i = s_i + l_i$ to $J_t = L_t + S_t$ to obtain the channel spin $i = j_i + J_t$. Then J is obtained by addition of the relative orbital angular momentum l,

$$J = i + l \tag{11.13}$$

When $l_i = 0$, we simply have $j_i = s_i$. It is not necessary to consider the coupling $J_t = L_t + S_t$ explicitly as will be apparent when this case is discussed below. In fact, it is not necessary that both L_t^2 and S_t^2 be constants of the motion.

If only one channel spin is possible, for instance, $J_t = 0$, then the shape of the angular distribution is uniquely determined without recourse to any knowledge of nuclear matrix elements. When more than one value is possible the angular distribution is essentially uniquely determined if it were possible to argue that all but one value of i make negligible contributions. Actually there is no physically plausible a priori reason for this. In the other two coupling schemes considered, the circumstances under which the angular distribution might be unique do correspond to reasonably well-understood physical approximations. Whether the approximations themselves are reasonable is another matter.

In j–j coupling we have the following scheme

$$s_i + l = j$$
$$j + J_t = J \tag{11.14}$$

We can think of the incident particle with its intrinsic angular momentum s_i moving about A with orbital angular momentum l so that, relative to A, its total angular momentum is j. This is then added (vectorially, of course) to the total angular momentum of A to form J.

Starting with given S_i, J_t, and J it is clear that, in general, several channel spins i and also several values of j may be possible. The observed result may happen to coincide with that corresponding to one of the values of j, say. This could then be interpreted as evidence for j–j coupling. More likely, this does not occur, and the result would be interpretable as a mixture of different j-values or different i-values.

In L–S coupling the scheme is as follows:

$$\mathbf{s}_i + \mathbf{S}_t = \mathbf{S}$$

$$\mathbf{l} + \mathbf{L}_t = \mathbf{L}$$

$$\mathbf{S} + \mathbf{L} = \mathbf{J} \tag{11.15}$$

Again the observations may fit an assignment of the S- and L-values of the compound state, in which case L–S coupling would work, or else a mixture could be called for.

In this discussion we have made reference to an angular distribution without discussion of what is observed. Of course, we refer to the correlation in propagation vectors of the incident particle X and whatever radiation is emitted in the decay of the compound nucleus. This could involve radiation emitted directly from the compound state, as in the case of elastic scattering, or radiation emitted after one or more transitions have taken place. An example of the latter would be the angular distribution of gamma rays emitted in inelastic scattering, $(p, p'\gamma)$ reactions say. Obviously, the angular distribution depends on the nature of this radiation as well as the quantum numbers involved in the decay of the compound nucleus. But it is unnecessary, at this stage, to be specific about this part of the problem. This is the advantage of the break-up of the problem of angular distributions as discussed in section 33. We have only to discuss the Λ-matrix for the formation of the compound nucleus. We know that it will have the form

$$\Lambda_{mm'} = \sum_{\nu} (-)^{m'} \, C(JJ\nu; m, -m') \, D^{\nu}_{m-m',0}(R) \, B_{\nu} \tag{11.16}$$

as in (10.24a). The parameter B_{ν}, which may contain a summation over orbital angular momenta, is obtained by inspection of the Λ-matrix; see section 33. Then, if

$$B_{\nu} = \sum_{ll'} b_{\nu}(ll')$$

whereas for gamma rays it was (section 33)

$$B_{\nu}(\gamma) = \sum_{ll'} b_{\nu}(\gamma; ll')$$

then we replace $b_{\nu}(\gamma; ll')$ in the γ–y correlation function with $b_{\nu}(ll')$ in the nuclear reaction where y designates the radiation emitted in the decay of the compound nucleus. Here $l(l')$ refer to the angular momentum of the gamma ray in $b_\nu(\gamma; ll')$.

As an example consider the reaction

$$B^{11} + p = C^{12*} = C^{12} + \gamma_2$$

The angular distribution of the gamma rays (γ_2) relative to the incident proton beam is now to be compared with the angular correlation between two gamma rays, γ_1 and γ_2, in the decay scheme

$$C^{12**} = C^{12*} + \gamma_1 = C^{12} + \gamma_1 + \gamma_2$$

where the double asterisk denotes an excited state of higher energy than the one referred to by the single asterisk. As a matter of definition we assume that C^{12*} represents the same excited state of C^{12} in both reactions so that the Λ-matrix characterizing the formation of this state differs only in the parameter B_ν in (11.16). The initial states in the two reactions represented above are different, but this is taken care of by the fact that the B_ν will differ for the proton capture and the emission of γ_1.

We first assume that the protons have a single orbital angular momentum value l and that this is the same as the angular momentum of the photon. Then for the proton

$$B_\nu(p) = b_\nu(p; ll) = g_\nu \, b_\nu(\gamma; ll) \qquad (11.17a)$$

and we have emphasized by the notation that $l = l'$. Equation (11.17a) defines the parameter g_ν. This parameter will depend on l, as well as on parameters defining the initial states. For the moment we suppress the explicit expression of this in the notation. If the γ_1–γ_2 correlation is

$$W_{\gamma_1\gamma_2} = \sum_\nu A_\nu(ll)P_\nu$$

the p–γ_2 correlation will be

$$W_{p\gamma_2} = \sum_\nu g_\nu \, A_\nu(ll)P_\nu \qquad (11.17b)$$

The fact that the initial states B^{11} (ground) and C^{12**} do not have the same angular momentum or parity is readily taken into account—see below. For the parity question we recall that for any l (>0) there is a multipole for each parity change and that the correlation function $W_{\gamma_1\gamma_2}$ is independent of the parity change.

For a mixed transition, orbital angular momenta l and l' contributing in the proton beam, we must have $|l - l'| = 2, 4, \cdots$. Although this is not the practical mixture in the case of gamma rays (there one has l-magnetic and $l+1$-electric or l-electric and $l+1$-magnetic radiations appearing together), the case of two photons, both electric or both magnetic, is easily treated with the same procedure as has been described already. The correlation function is symmetrized as in (10.51) except that $W_{ll'}$ is obtained from (10.27) by replacing F_ν in (10.27b) by f_ν as given in (10.27c). This defines the mixed γ–γ correlation for any pair of angular momenta l and l'. It also defines the parameter $b_\nu(\gamma; ll')$. If again we define $g_\nu(ll')$ by

$$b_\nu(ll') = g_\nu(ll')\, b_\nu(\gamma; ll')$$

then the p–γ_2 angular correlation for a mixed proton capture is

$$W_{p\gamma_2} = \sum_\nu \sum_{ll'} g_\nu(ll')\, A_\nu(ll') P_\nu \qquad (11.18)$$

where $A_\nu(ll')$ are the coefficients of P_ν in the mixed γ_1–γ_2 angular correlation.

The only conceivable case which is not included in the foregoing procedure is the case of a mixture where one of l or l' is zero. The pure case $l = 0$, which does not correspond to any gamma ray, gives isotropy as expected. For the mixed case, as in s–d interference, we need to replace $C(ll'\nu; 1, -1)$ by $C(ll'\nu; 00)$ to convert from a γ–γ to an α–γ correlation. This is also done in defining the g_ν which would give the ratio of $b_\nu(ll')$ to $b_\nu(\alpha; ll')$; see section 34.

From the results of section 33, the parameter B_ν for gamma radiation is

$$B_\nu(\gamma) = W(JJLL; \nu J_1)\, C(LL\nu; 1, -1) \qquad (11.19)$$

when we consider a pure multipole of angular momentum L. We have omitted certain irrelevant multiplicative factors. The angular momenta of initial and final states are J_1 and J, respectively, so that

$$\mathbf{J}_1 = \mathbf{L} + \mathbf{J}$$

We now compare this with channel spin formulation of the compound state. Instead of the initial state J_1 breaking up into a final state (J) plus radiation L, we form state J from a state with angular momentum i absorbing angular momentum l. The angular momentum l is associated with orbital motion of a non-relativistic particle. Hence there are two changes which it is necessary to incorporate in (11.19). First $C(LL\nu; 1, -1)$ is replaced by $C(ll\nu; 00)$. Second, J_1 is replaced by i; also, we change the notation $L \to l$ throughout. Then we have

$$B_\nu(p) = W(JJll; \nu i)\, C(ll\nu; 00) \qquad (11.20)$$

This result should be inserted in (11.16) to obtain the angular distribution in a p–y reaction for a particular channel spin

$$W = \sum_{\nu} C(ll\nu; 00)\ W(JJll; \nu i)\ B_{\nu}(y)\ P_{\nu}(\cos \theta) \qquad (11.20')$$

If this angular distribution obtained with a fixed channel spin i is denoted by $W(i)$, then the general case is

$$W = \sum_{i} k_i\ W(i) \qquad (11.21)$$

where k_i are certain coefficients which depend on a more intimate knowledge of nuclear forces than we now have. These are generally treated as parameters to be adjusted from the experimental data. The values of i are those consistent with the triangular relation $\Delta(Jli)$.

Two comments may now be made. In obtaining the result given above we quite properly ignored any distinction between emission and absorption. This distinction would lie in replacing

$$(J_f m_f | T_{LM} | J_i m_i) \sim C(J_i L J_f; m_i M)\ \delta_{m_f, m_i + M}$$

where now subscripts f and i mean final and initial state, by

$$(J_i m_i | T_{LM}^+ | J_f m_f) \sim (-)^{L-M}\ C(J_f L J_i; m_f, -M)\ \delta_{m_i, m_f - M}$$

The projection quantum number dependence for these two is just the same, as is seen when (3.17a) is applied.[11] The second point is that the contributions of the different channel spins add incoherently. This is obvious when we consider that i replaces J_1, the initial state in the comparison problem.

The next subject for consideration is the j–j coupled nuclear reaction. We can construct the required answer by comparing this coupling scheme with the channel spin representation. To repeat, in the latter case we construct J out of s_i and J_t and l as follows:

$$\mathbf{i} = \mathbf{s}_i + \mathbf{J}_t$$

$$\mathbf{J} = \mathbf{l} + \mathbf{i}$$

In j–j coupling the same vectors produce the same resultant by recoupling in the manner shown in (11.14). By definition the recoupling is accomplished by

$$R_{ji} = [(2j + 1)(2i + 1)]^{\frac{1}{2}}\ W(s_i l J_t J; ji) \qquad (11.22)$$

[11] We have used $T_{LM}^+ = (-)^{L-M} T_{L,-M}$. The phase in these irreducible tensors may always be chosen so that this result applies; see R. Huby, *Proc. Phys. Soc.* **A 67**, 1103 (1954). That $T_{LM}^+ \sim (-)^M T_{L,-M}$ is always necessary follows from the definition (5.1).

Thus, it is possible to express the angular distribution in j–j coupling in terms of the pure channel spin case. In other words, to within a multiplicative factor, the assumption of a coupling scheme fixes the constants k_i. In fact, the angular distribution is

$$W(jj) = \sum_i (2i + 1)\, W^2(s_i l J_t J; ji)\, W(i) \tag{11.23}$$

where $W(i)$ is defined in (11.20′). Using this result and, for definiteness, assuming that the compound nucleus decays by emitting a gamma ray of angular momentum L_2, leaving the nucleus in the state J_2, the angular distribution is

$$W(jj) = (2l + 1)(2j + 1) \sum_\nu C(ll\nu; 00)\, W(lljj; \nu s_i)$$

$$\times\, W(JJjj; \nu J_t)\, F_\nu(L_2 J_2 J)\, P_\nu(\cos\theta) \tag{11.24}$$

where F_ν is defined in (10.27b). For $s_i = \frac{1}{2}$ (neutrons, protons, H^3, He^3) this reduces to

$$W(jj) = (2j + 1) \sum_\nu C(jj\nu; \tfrac{1}{2} - \tfrac{1}{2})\, W(JJjj; \nu J_t)\, F_\nu(L_2 J_2 J) P_\nu \tag{11.24′}$$

by an easily established identity.[12] For $j = \frac{1}{2}$ we have $\nu = 0$ and 1 only. But, since ν must be even, only $\nu = 0$ enters, and the angular distribution is isotropic.

The result (11.24) is obtained by using certain identities in the Racah coefficients. It can be obtained alternatively as follows. For particles of spin s_i a plane wave representation is expanded into wave functions of the total angular momentum:

$$\chi_{s_i\mu}\, e^{i\mathbf{k}\cdot\mathbf{r}} = 4\pi \sum_j \sum_{lm} i^l\, j_l(kr)\, Y^*_{l,m-\mu}\, C(s_i lj; \mu, m-\mu)\, \psi_{jm}$$

The diagonal element of the Λ-matrix when $\mathbf{f}$ is along the quantization axis is

$$\Lambda_M(jj) = \sum_{mM_t} |\, C(s_i lj; m0)\, (jm|J_t M_t|JM)\, |^2$$

$$\sim \sum_{mM_t} [C(s_i lj; m0)\, C(jJ_t J; mM_t)]^2\, \delta_{m+M_t, M}$$

Evaluation of $\Lambda_M(jj)$ along conventional lines [by recoupling according to (6.4b)] leads to (11.24).

In an analogous manner we can obtain corresponding results in L–S coupling. The initial state consists in part of a plane wave for the in-

---

[12] The identity is written by equating (11.24) and (11.24′) and verified by use of the results given in Appendix I and (3.19).

cident particles which is expanded into waves of definite orbital angular momentum. As before, we select one of these, making the assumption that all others either contribute nothing because of angular momentum conservation or contribute very ineffectively (e.g., the effect of the potential barrier could serve to eliminate all but the smallest possible orbital angular momentum). The orbital function is, of course, $\psi_{lm_l} \sim Y_{lm_l}$. The remaining parts of the initial state are the intrinsic spin function for the incident particles $\chi_{s_im_i}$ and the target ground state function $\Psi_{J_tM_t}$. Now

$$\Psi_{J_tM_t} = \sum_{M_L} C(S_tL_tJ_t; M_t-M_L, M_L)\, \psi_{S_t,M_t-M_L}\, \psi_{L_tM_L}$$

is a decomposition of the target function into wave functions of the target orbital and spin angular momentum. So, for the initial state we have altogether

$$\Psi_{in} = \psi_{lm_l} \chi_{s_im_i} \sum_{M_L} C(S_tL_tJ_t; M_t-M_L, M_L)\, \psi_{S_t,M_t-M_L}\, \psi_{L_tM_L}$$

Then we combine ψ_{lm_l} and $\psi_{L_tM_L}$ as follows:

$$\psi_{lm_l}\, \psi_{L_tM_L} = \sum_L C(lL_tL; m_lM_L)\, \Psi_{L,m_l+M_L}$$

Similarly

$$\chi_{s_im_i} \psi_{S_t\, M_t-M_L} = \sum_S C(s_iS_tS; m_i, M_t-M_L)\, \Psi_{S\ m_i+M_t-M_L}$$

Finally, we combine S and L to form J according to

$$\Psi_{S\ m_i+M_t-M_L}\, \Psi_{L\ m_l+M_L}$$
$$= \sum_J C(SLJ; m_i+M_t-M_L, m_l+M_L)\, \Psi_{JM}\, \delta_{m_i+M_t+m_l,M}.$$

Thus

$$(\Psi_{JM}, \Psi_{in}) = \sum_{LS} \sum_{M_L} C(S_tL_tJ_t; M_t-M_L, M_L)\, C(lL_tL; m_lM_L)$$
$$\times C(s_iS_tS; m_i, M_t-M_L)\, C(SLJ; m_i+M_t-M_L, m_l+M_L)$$
$$\times \delta_{m_i+M_t+m_l,M} \quad (11.25)$$

We can set $m_l = 0$ corresponding to a choice of quantization axis along $\mathbf{l}$. For fixed M we have only one independent projection quantum number $m_i = M - M_t$. Then the scalar product (11.25) must be squared and averaged over m_i (or M_t). Thus, for a particular L and S,

$$\Lambda_{MM}(LS) \sim \sum_{m_i} \left[\sum_{M_L} C(S_tL_tJ_t; M-m_i-M_L, M_L)\, C(lL_tL; 0M_L) \right.$$
$$\left. \times C(s_iS_tS; m_i, M-m_i-M_L)\, C(SLJ; M-M_L, M_L) \right]^2 \quad (11.26)$$

This sum can be evaluated by the techniques used so often before. Rather than go through each step in detail we shall give the result and then attempt to see, in an alternative way, how it comes about. First, the result can be expressed in terms of the channel spin distribution $W(i)$. Then this again amounts to an evaluation of the coefficients k_i.

$$W(LS) = \sum_i (2i + 1)(2J_t + 1)(2L + 1)(2S + 1) \, W(i)$$
$$\times \, W^2(lL_tJS; Li) \, W^2(s_iS_tiL; SJ_t) \quad (11.27)$$

When there is interference between different values of l, L, S, and J, the changes to be made are as follows:

1. Change $(2L + 1)$ into $[(2L + 1)(2L' + 1)]^{\frac{1}{2}}$.
2. Change $(2S + 1)$ into $[(2S + 1)(2S' + 1)]^{\frac{1}{2}}$.
3. Change the square of the first Racah coefficient into the product of two Racah coefficients where one is just what appears in (11.27) and in the other l, J, S, and L are replaced by l', J', S', and L'—the interfering values. Similarly, in the last factor two different Racah coefficients appear instead of a square of one. They differ by the replacement $S \to S'$, $L \to L'$. Then we sum over all interfering angular momenta.

There is another way in which this result may be expressed. This is seen from the fact that where in j–j coupling we have

$$\mathbf{j} = \mathbf{l} + \mathbf{s}_i$$
$$\mathbf{J}_t = \mathbf{L}_t + \mathbf{S}_t$$
$$\mathbf{J} = \mathbf{j} + \mathbf{J}_t \quad (11.28a)$$

in L–S coupling this is changed to

$$\mathbf{S} = \mathbf{s}_i + \mathbf{S}_t$$
$$\mathbf{L} = \mathbf{l} + \mathbf{L}_t$$
$$\mathbf{J} = \mathbf{S} + \mathbf{L} \quad (11.28b)$$

Now to recouple from the j–j to the L–S scheme requires, as we have already seen in section 35, a unitary matrix whose elements are given in terms of X-coefficients. Since there are two matrix elements, entering in the angular distribution function, which correspond to the formation of the compound state, we expect two X-coefficients to enter. The X-coefficients will be the same if no interferences are present. Considering this case for simplicity of writing, the particle–gamma angular distribution is (for $s_i = \frac{1}{2}$)

$$W(LS) \sim \sum_\nu \eta_\nu \Gamma_\nu^2 \, F_\nu(L_2 J_2 J) \, P_\nu(\cos \theta) \quad (11.29)$$

where, as given by Satchler,[11]

$$\eta_\nu = (2J + 1)^{\frac{1}{2}}(2j + 1)(-)^{J_t - J - \frac{1}{2}} C(\nu jj; 0\tfrac{1}{2}) W(JJjj; \nu J_t) \quad (11.29a)$$

$$\Gamma_\nu = (2j + 1)^{\frac{1}{2}} X(J_t L_t S_t; jl\tfrac{1}{2}; JLS) \quad (11.29b)$$

Here η_ν will be recognized as the parameter that appears in (11.24') in the j–j coupling case.

This result can be obtained almost directly from (11.26) if we utilize the definition of the X-coefficient in terms of C-coefficients. From section 35 we have

$$\Phi_{JM}(LS) = \sum_{j_1 j_2} T(j_1 j_2; LS)\, \Psi_{JM}(j_1 j_2)$$

or

$$\sum_{\tau m \mu} C(LSJ; m, M-m)\, C(l_1 l_2 L; \mu, m-\mu)\, C(s_1 s_2 S; \tau, M-m-\tau)$$

$$\times \chi_{s_1 \tau}\, \chi_{s_2\, M-m-\tau}\, \psi_{l_1 \mu_1}\, \psi_{l_2\, m-\mu}$$

$$= \sum_{j_1 j_2} T(j_1 j_2; LS) \sum_{m_1}\sum_{\tau_1} C(j_1 j_2 J; m_1, M-m_1)\, C(l_1 s_1 j_1; m_1 - \tau_1, \tau_1)$$

$$\times \sum_{\tau_2} C(l_2 s_2 j_2; M-m_1-\tau_2, \tau_2)\, \chi_{s_1 \tau_1}\, \chi_{s_2 \tau_2}\, \psi_{l_1\, m_1 - \tau_1}\, \psi_{l_2\, M-m_1-\tau_2}$$

where we have substituted from (11.3) and (11.4) into (11.5). We now take the scalar product of this equation with $\psi_{l_1 \mu_1}\, \psi_{l_2 \mu_2}\, \chi_{s_1 \sigma_1}\, \chi_{s_2 \sigma_2}$. Each of these four functions, it must be remembered, is in a separate space. From their orthonormality we obtain immediately

$$\mu_1 = \mu, \qquad m = \mu_1 + \mu_2$$

$$\tau = \sigma_1, \qquad M = \mu_1 + \mu_2 + \sigma_1 + \sigma_2$$

and

$$\sigma_1 = \tau_1, \qquad \sigma_2 = \tau_2$$

$$m_1 = \mu_1 + \sigma_1, \qquad M = \mu_1 + \mu_2 + \sigma_1 + \sigma_2$$

Therefore we obtain

$$C(LSJ; \mu_1 + \mu_2, M - \mu_1 - \mu_2)\, C(l_1 l_2 L; \mu_1 \mu_2)\, C(s_1 s_2 S; \sigma_1, M - \mu_1 - \mu_2 - \sigma_1)$$

$$= \sum_{j_1 j_2} T(j_1 j_2; LS)\, C(j_1 j_2 J; \mu_1 + \sigma_1, M - \mu_1 - \sigma_1)$$

$$\times C(l_1 s_1 j_1; \mu_1 \sigma_1)\, C(l_2 s_2 j_2; \mu_2, M - \mu_1 - \mu_2 - \sigma_1) \quad (11.30)$$

where σ_2 has been eliminated. Next multiply this equation by $C(l_2 s_2 j_2'; \mu_2, M - \mu_1 - \mu_2 - \sigma_1)$ and sum over μ_2. On the right side only the last C-coefficient depends on μ_2, so that the μ_2 sum is simply $\delta_{j_2 j_2'}$.

Hence the j_2 sum can be done by replacing j_2 by j_2' in the remaining terms on the right-hand side. Dropping the prime on j_2 and recalling the definition of $T(j_1 j_2; LS)$ in terms of the X-coefficient, we find

$$\sum_{\mu_2} C(LSJ; \mu_1+\mu_2, M-\mu_1-\mu_2)\, C(l_1 l_2 L; \mu_1\mu_2)$$

$$\times\, C(s_1 s_2 S; \sigma_1, M-\mu_1-\mu_2-\sigma_1)\, C(l_2 s_2 j_2; \mu_2, M-\mu_1-\mu_2-\sigma_1)$$

$$= \sum_{j_1} [(2j_1+1)(2j_2+1)(2L+1)(2S+1)]^{\frac{1}{2}}\, X(l_1 s_1 j_1; l_2 s_2; LSJ)$$

$$\times\, C(j_1 j_2 J; \mu_1+\sigma_1, M-\mu_1-\sigma_1)\, C(l_1 s_1 j_1; \mu_1\sigma_1) \quad (11.31)$$

The relation (11.31) is just the form required to evaluate the sum in (11.26) [13]: i.e., a single sum over a projection quantum number of the product of four C-coefficients. In fact, in (11.26), if we write the last C-coefficient first and then make the transposition (3.16b) in this coefficient as well as in the one that appears first in (11.26), we find that the M_L sum is the left-hand side of (11.31) term by term (with an appropriate transcription of notation) and that expression (11.31) is equal to

$$(-)^{S+L-J}(-)^{St+Lt-Jt} \sum_{j} [(2j+1)(2J_t+1)(2L+1)(2S+1)]^{\frac{1}{2}}$$

$$\times\, X(ls_i j; L_t S_t J_t; LSJ)\, C(jJ_t J; m_i, M-m_i)\, C(ls_i j; 0m_i)$$

It is now a comparatively simple matter to carry out the m_i sum in (11.26) after squaring the expression given above. The result leads to (11.29) when the same restrictions are made and, more generally, is equivalent to (11.27).

When there are interferences we need only symmetrize; i.e., if there are interferences between j and j', then we replace $\varphi(j)$ by $[\varphi(j)\,\varphi(j')]^{\frac{1}{2}}$ where φ is any function of j. Then we sum over j and j' with arbitrary (symmetrized) coefficients $C_{jj'} = C_{j'j}$.

For the details of the comparison of calculated results with observations in the case of p-shell (mass number between 4 and 16) reactions, the paper of Christy [10] should be consulted. The conclusion of that paper is somewhat indecisive, insofar as no evidence can be seen in favor of either extreme coupling scheme: j-j or L-S. Subsequent investigations of the properties of nuclei in the p-shell indicate that a form of intermediate coupling may be more appropriate.[14]

[13] Equations (11.30) and (11.31) should be compared with the analogous equation (6.4) for the Racah coefficient, or R-matrix.

[14] For example, A. M. Lane, *Proc. Phys. Soc.* **A 66**, 977 (1953).

XII. IDENTICAL PARTICLES

37. IDENTICAL PARTICLES IN j–j COUPLING

Whenever the physical structure under consideration consists of two or more identical particles, the symmetry requirements imposed by the statistics must be taken into account. The problem is the construction of a wave function Ψ for the total system which is an eigenfunction of the total angular momentum and which also fulfills the symmetry condition

$$P_{ij}\Psi = \pm\Psi \qquad (12.1)$$

where P_{ij} is the permutation operator for any pair of identical particles i and j and the $(+)$ $(-)$ sign goes with (Bose) (Fermi) statistics.

For two particles the problem is very simple. $\Psi(1, 2)$ is an appropriate non-symmetrized wave function; $\Psi(1, 2) \pm \Psi(2, 1)$ is the properly symmetrized function. When $\Psi(1, 2)$ is expressed in terms of single particle orbitals $\varphi_A(1)$, $\varphi_B(2)$ etc.,

$$\Psi = \frac{1}{\sqrt{2}}\,[\varphi_A(1)\,\varphi_B(2) \pm \varphi_A(2)\,\varphi_B(1)] \qquad (12.2)$$

In this case we distinguish between equivalent and non-equivalent orbitals. Equivalent orbitals are those for which φ_A and φ_B are identical as far as the angular momentum quantum numbers and the radial part of the single-particle wave functions are concerned. In the j–j coupling model this would mean the same j and l for each particle, and we would designate the wave function by φ_{jm} (section 30). Then φ_A and φ_B can differ only in m. For Fermi particles they must differ in m. This is the case of greatest interest since we are interested in nuclear structure with more than one particle outside closed shells.

Other systems, obeying Bose statistics, offer no difficulty. For instance, for the almost trivial case of two alpha particles, the resultant orbital angular momentum L must be even. The case of two deuterons is only slightly more complicated. We consider a system like the D_2-molecule, so that consideration need be given only to the symmetry between the two deuterons, and that between the two protons (or two neutrons) which belong to the separate deuterons is irrelevant. If j_1 and j_2 denote the deuteron angular momenta, the spin function of the molecule is

$$\Psi_{JM_J} = \sum_{M_1} C(j_1 j_2 J; M_1, M_J - M_1) \, \psi_{j_1 M_1}(1) \, \psi_{j_2 \, M_J - M_1}(2) \quad (12.3)$$

where $\mathbf{J} = \mathbf{j}_1 + \mathbf{j}_2$ is the spin operator for the entire molecule. Here $\psi_{j_1 M_1}$ and $\psi_{j_2 M_2}$ depend on the internal coordinates of each deuteron. Since $j_1 = j_2 = j = 1$, interchanging the two deuterons has the following effect:

$$P_{12} \, \Psi_{JM_J} = (-)^{2j-J} \, \Psi_{JM_J} = (-)^J \, \Psi_{JM_J} \quad (12.4)$$

as may be seen by applying (3.16b). The total wave function of the molecule is the product of Ψ_{JM_J} and a rotational eigenfunction. One could (and should) also introduce a wave function for the molecular vibration. However, this is a function of r, the distance between the centers of gravity of the two deuterons, and is always symmetric. Therefore we restrict our attention to the spin-rotation space and write

$$\Psi = Y_{LM_L} \Psi_{JM_J}$$

because the relative orbital motion is not coupled to the internal degrees of freedom. It follows that

$$P_{12}\Psi = (-)^{L+J}\Psi \quad (12.5)$$

so that $L + J$ = even. Therefore we have two possibilities: (1) $J = 0$ or 2 and L is even, (2) $J = 1$ and L is odd. In D_2 the levels with L even correspond to the orthodeuterium molecule and L odd is characteristic of paradeuterium. The generalization to any other system of the Bose particles is immediate.

For two Fermi particles of angular momenta j_1 and j_2 the wave function in non-equivalent orbits is

$$\Psi_{JM} = \frac{1}{\sqrt{2}}$$

$$\times \sum_m C(j_1 j_2 J; m, M-m)[\varphi_{j_1 m}(1) \, \varphi_{j_2 \, M-m}(2) - \varphi_{j_1 m}(2) \, \varphi_{j_2 \, M-m}(1)]$$

$$(12.6a)$$

If $j_1 = j_2 = j$, we can write $m' = M - m$ in the second sum, and then, dropping the prime,

$$\Psi_{JM} = \tfrac{1}{2}[1 - (-)^{2j-J}] \sum_m C(jjJ; m, M-m) \, \varphi_{jm}(1) \, \varphi_{j \, M-m}(2) \quad (12.6b)$$

where again (3.16b) has been used and the factor in front has been adjusted to give a Ψ_{JM} normalized to unity. It is seen that the symmetry requirement $(P_{12}\Psi = -\Psi)$ is

$$(-)^{2j-J} = -1$$

or, since j is of the form $n + \frac{1}{2}$ and J is an integer,

$$(-)^J = +1$$

and J must be even.

For three or more equivalent particles these considerations become considerably more involved, and we present only a brief discussion of the problem.[1] Consider three Fermi particles in j–j coupling. Then, aside from normalization constants, a possible wave function for which the total angular momentum and projection quantum number are fixed is

$$\Psi(j^2(J')jJ) = \sum_m C(J'jJ; m, M-m)$$

$$\times \left[\sum_{m'} C(jjJ'; m', m-m') \; \varphi_{jm'}(1) \; \varphi_{j\,m-m'}(2) \; \varphi_{j\,M-m}(3) \right] \quad (12.7)$$

Of course, Ψ must be antisymmetrical relative to the operators P_{12}, P_{23}, P_{31}. The quantity inside the brackets in (12.7) is antisymmetrical relative to P_{12} if J' is even. But, in general, there is more than one value of J' permitted. Of course, we do not know at this stage which J-values will be permitted, but, in general, the possible values of J' are the even integers which are common to the overlap region defined by the following two conditions:

$$|J - j| \leq J' \leq J + j$$

$$0 \leq J' \leq 2j - 1$$

$J' = 0$ only if $J = j$. As an example, for $j = \frac{5}{2}$ and $J = \frac{9}{2}$ we can have $J' = 2$ and 4. Hence, the proper wave function Ψ_{JM} must be a linear combination of these two. More generally,

$$\Psi_{JM} = \sum_{J'} F_{J'} \Psi(j^2(J')jJ) \quad (12.8)$$

and the coefficients $F_{J'}$ are commonly known as coefficients of fractional parentage.[2] The customary notation, for three particles, is

$$F_{J'} \equiv (j^2(J')jJ \,|\, \}j^3J)$$

[1] A much more complete treatment is given by M. Redlich, Ph.D. dissertation, On the Agreement between Nuclear Shell Theory and Experiment, Princeton University, January 1954. See also B. H. Flowers, *Proc. Roy. Soc. London* **A 212,** 248 (1952); A. R. Edmonds and B. H. Flowers, *Proc. Roy. Soc. London* **A 214,** 515; **A 215,** 120 (1952).

[2] The usefulness of the concept of fractional parentage in nuclear reaction phenomena has been emphasized by A. M. Lane and D. H. Wilkinson, *Phys. Rev.* **97,** 1199 (1955).

The additional freedom introduced in (12.8) is just what is needed for the additional requirement $P_{23}\Psi = -\Psi$. If this is fulfilled then, of course, $P_{31}\,\Psi_{JM} = -\Psi_{JM}$ since $P_{31} = P_{12}P_{23}P_{12}$. This additional condition together with the normalization condition

$$\sum_{J'} (F_{J'})^2 = 1 \tag{12.9}$$

determines the coefficients $F_{J'}$ for permissible J-values—within an irrelevant phase.[3]

Applying the antisymmetrization condition, we see that

$$\sum_{J'} F_{J'} \sum_{mm'} C(J'jJ; m, M-m)\, C(jjJ'; m', m-m')$$

$$\times \varphi_{jm'}(1)\, \varphi_{j\,m-m'}(2)\, \varphi_{j\,M-m}(3)$$

$$= -\sum_{J''} F_{J''} \sum_{\mu\mu'} C(J''jJ; \mu, M-\mu)\, C(jjJ''; \mu', \mu-\mu')$$

$$\times \varphi_{j\mu'}(1)\, \varphi_{j\,\mu-\mu'}(3)\, \varphi_{j\,M-\mu}(2) \tag{12.10}$$

We now take the scalar product of both sides of (12.10) with

$$C(jj\lambda; \bar{m}', \bar{m}-\bar{m}')\, C(\lambda jJ; \bar{m}, M-\bar{m})\, \varphi_{j\bar{m}'}(1)\, \varphi_{j\,\bar{m}-\bar{m}'}(2)\, \varphi_{j\,M-\bar{m}}(3)$$

$$\tag{12.11}$$

and sum over $\bar{m}'$ and $\bar{m}$. This gives, after dropping the bars and changing λ to J',

$$F_{J'} = -\sum_{J''} \sum_{\mu\mu'} C(J''jJ; \mu, M-\mu)\, C(jjJ''; \mu', \mu-\mu')$$

$$\times C(jjJ'; \mu', M-\mu)\, C(J'jJ; M-\mu+\mu', \mu-\mu')\, F_{J''} \tag{12.12}$$

If we now compare the μ, μ' sum with (6.6a), we see that the two sums are identical (within a phase) if we first make the transcription of notation $j_1,j_2,j_3,j',j'',j,\mu_1,\mu_2,\mu_3 = j,j,j,J',J'',J,M-\mu,\mu',\mu-\mu'$, respectively, and interchange the first pair of angular momenta on the first and third C-coefficients above. This gives

$$F_{J'} = (-)^{3j-J+1} \sum_{J''} (-)^{J''-J'} R_{J''J'}\, F_{J''} \tag{12.13}$$

---

[3] From the discussion that follows it will be clear that there is one arbitrary phase in the set of coefficients $F_{J'}$ and that, if one of these coefficients is taken to be real, then all of them will be real. We do adopt the convention that all fractional parentage coefficients are real.

with J' and J'' even integers. Since $4j$ is an even integer, we can rewrite (12.13) in the form [4]

$$\sum_{J''} F_{J''}(A_{J''J'} - \delta_{J''J'}) = 0$$

$$A_{J''J'} = (-)^{J-j} R_{J''J'}$$
$$= (-)^{J-j}[(2J'+1)(2J''+1)]^{\frac{1}{2}} W(jjjJ; J'J'') \quad (12.14)$$

It follows then that the permissible values of J are those for which the determinantal equation

$$\det (A_{J''J'} - \delta_{J''J'}) = 0 \tag{12.15}$$

is satisfied; that is, the matrix A must have eigenvalues $+1$, and only for these cases do we get an antisymmetrical wave function.

Taking up the configurations j^3 in turn, for $j = \frac{1}{2}$ no possible value of J fulfills (12.15). This is hardly surprising since for equivalent particles only $2j + 1 = 2$ states are available. For $j = \frac{3}{2}$, we need to consider $J = \frac{3}{2}$ which requires $J', J'' = 0, 2$ or $J = \frac{1}{2}, \frac{5}{2}$ or $\frac{7}{2}$ and $J', J'' = 2$ only. We consider $J = \frac{3}{2}$ first. The pertinent Racah coefficients are

$$W(\tfrac{3}{2}\tfrac{3}{2}\tfrac{3}{2}\tfrac{3}{2}; 0J'') = W(\tfrac{3}{2}\tfrac{3}{2}\tfrac{3}{2}\tfrac{3}{2}; J'0) = -\tfrac{1}{4}, \qquad W(\tfrac{3}{2}\tfrac{3}{2}\tfrac{3}{2}\tfrac{3}{2}; 22) = \tfrac{3}{20}$$

It is clear that $J = \frac{3}{2}$ is a permissible value of the total angular momentum.

For $J \neq j$ we use

$$W(\tfrac{3}{2}\tfrac{3}{2}\tfrac{3}{2}\tfrac{1}{2}; 22) = \tfrac{1}{10}$$
$$W(\tfrac{3}{2}\tfrac{3}{2}\tfrac{3}{2}\tfrac{5}{2}; 22) = \tfrac{1}{10}$$
$$W(\tfrac{3}{2}\tfrac{3}{2}\tfrac{3}{2}\tfrac{7}{2}; 22) = -\tfrac{1}{10}$$

so that $A_{22} = \pm\frac{1}{2}$.

Quite generally, for the configuration j^3 it can be seen that J cannot have either of the values $3j - 1$ or $3j - 2$. For either of these values it is necessary that both J' and J'' be restricted to the single value $2j - 1$. This is the only even integer for which the triangle rules hold. From (6.11) and (12.14) we find

$$A_{2j-1,2j-1} = -\tfrac{1}{2}$$

for both $J = 3j - 1$ and $J = 3j - 2$. We can also see that $J = \frac{1}{2}$ is excluded for all j^3 configurations. In this case J' and J'' are again restricted to a single value, namely $j \pm \frac{1}{2}$, whichever is even. In both cases the Racah coefficient in (12.14) is obtained from (6.11), and in both cases we find $A_{J'J'} = -\frac{1}{2}$.

[4] This equation is derived in a paper by C. Schwartz and A. De-Shalit, *Phys. Rev.* **94**, 1257 (1954).

Returning to the configuration j^3, it is seen that $J = \frac{3}{2}$ is the only permissible value. The normalized wave function is given by (12.8) with the following fractional parentage coefficients:

$$F_0 = \sqrt{\tfrac{1}{6}}, \qquad F_2 = -\sqrt{\tfrac{5}{6}} \qquad (12.16)$$

We now consider the configuration $(\frac{5}{2})^3$. Then $J', J'' = 0, 2, 4$. Hence, the only possible values of J available for consideration are $J = \frac{1}{2}, \frac{3}{2}, \frac{5}{2}, \frac{7}{2}, \frac{9}{2}$. Again $J', J'' = 0$ occurs only for $J = j = \frac{5}{2}$.

Taking up each J-value in turn, we first consider $J = \frac{1}{2}$. Then $J' = J'' = 2$ is the only possibility. But

$$A_{22} = -\tfrac{1}{2}$$

which excludes $J = \frac{1}{2}$.

For $J = \frac{3}{2}$ we need $J', J'' = 2, 4$

$$A_{22} = \tfrac{4}{7}, \qquad A_{24} = A_{42} = -\tfrac{3}{7}\sqrt{\tfrac{5}{2}}, \qquad A_{44} = -\tfrac{9}{126}$$

and (12.15) is satisfied. Thus $J = \frac{3}{2}$ is an acceptable value.

For $J = \frac{5}{2}$ we use

$$A_{0J'} = A_{J'0} = -(2J' + 1)^{\frac{1}{2}}/6,$$

$$A_{22} = -\tfrac{1}{12}, \qquad A_{24} = A_{42} = (\tfrac{5}{4})^{\frac{1}{2}}, \qquad A_{44} = \tfrac{1}{4}$$

and again (12.15) is fulfilled so that $J = \frac{5}{2}$ is permissible.

For $J = \frac{7}{2}$ we find

$$A_{22} = -\tfrac{1}{2}, \qquad A_{44} = -\tfrac{1}{2}, \qquad A_{24} = A_{42} = 0$$

The determinantal condition (12.15) is not fulfilled, and $J = \frac{7}{2}$ is excluded.

Finally, for $J = \frac{9}{2}$ we have

$$A_{22} = -\tfrac{5}{28}, \qquad A_{24} = A_{42} = -\tfrac{3}{28}\sqrt{33}, \qquad A_{44} = 9 \times \tfrac{19}{252}$$

so that $J = \frac{9}{2}$ satisfies (12.15).

Summarizing, $J = \frac{3}{2}, \frac{5}{2}$, and $\frac{9}{2}$ are the only possible J-values for the configuration $(\frac{5}{2})^3$. The wave functions are obtained from the fractional parentage coefficients given in Table 1. In Table 1 we have arbitrarily

Table 1. Fractional Parentage Coefficients for the Configuration $(\frac{5}{2})^3$

J	F_0	F_2	F_4
$\frac{3}{2}$		$\sqrt{\tfrac{5}{7}}$	$-\sqrt{\tfrac{2}{7}}$
$\frac{5}{2}$	$\tfrac{1}{3}\sqrt{2}$	$-\tfrac{1}{3}\sqrt{\tfrac{5}{2}}$	$-\sqrt{\tfrac{1}{2}}$
$\frac{9}{2}$		$\sqrt{\tfrac{3}{14}}$	$-\sqrt{\tfrac{11}{14}}$

chosen $F_4 < 0$. It is clear that, as we proceed to higher values of j or when the number of equivalent particles increases, these procedures become quite laborious. The more powerful and elegant methods of Flowers and Edmonds [1] should then be utilized.

As an application of the results found above we may consider the magnetic dipole or electric quadrupole moment calculated with these wave functions. We may compare the results with the extreme single-particle model where only a single odd particle is considered.

Consider the expectation value of any operator in the form of a sum of single-particle operators. Thus, for three particles

$$\langle \Omega \rangle = (\Psi_{JM} | \Omega(1) + \Omega(2) + \Omega(3) | \psi_{JM}) \qquad (12.17)$$

and we shall eventually take $M = J$. Obviously, $\Omega(1)$ operates in space 1 and is the unit operator in spaces 2 and 3, etc. Then, using (12.8) for the wave function,

$$\langle \Omega \rangle = \sum_{J'J''} F_{J'} F_{J''} \sum_{i=1}^{3} (j^2(J')jJ | \Omega(i) | j^2(J'')jJ) \qquad (12.18a)$$

For example,

$$(j^2(J')jJ | \Omega(1) | j^2(J'')jJ) = \sum_{mm'} \sum_{\mu\mu'} C(J'jJ; m, M-m)$$

$$\times C(jjJ'; m', m-m') C(J''jJ; \mu, M-\mu) C(jjJ''; \mu', \mu-\mu')$$

$$\times (jm' | \Omega(1) | j\mu') \delta_{\mu-\mu',m-m'} \delta_{M-m,M-\mu} \qquad (12.18b)$$

Here m' and μ' are used as projection quantum numbers for particle number 1. Using the Wigner–Eckart theorem, we find

$$(j^2(J')jJ | \Omega(1) | j^2(J'')jJ)$$

$$= (j \| \Omega(1) \| j) \sum_{m} C(J'jJ; m, M-m) C(J''jJ; m, M-m)$$

$$\times \sum_{m'} C(jjJ'; m', m-m') C(jjJ''; m', m-m') C(j\lambda j; m'0) \qquad (12.19)$$

where λ is the rank of the tensor $\Omega(1)$. The sum over m' (with m held fixed) is carried out in the now familiar way. Note that only three C-coefficients are involved. The result for the sum of the product of these three coefficients is

$$\sum_{m'} = (-)^{\lambda}[(2j+1)(2J''+1)]^{\frac{1}{2}} C(\lambda J''J'; 0m) W(\lambda jJ'j; jJ'')$$

When this is re-inserted in (12.19) and the sum over m is carried out in the same way, we get

$$(j^2(J')jJ | \Omega(1) | j^2(J'')jJ) = (J \| \Omega(1) \| J) C(J\lambda J; M0)$$

which is the Wigner–Eckart theorem, of course, and

$$(J \parallel \Omega(1) \parallel J)/(j \parallel \Omega(1) \parallel j)$$

$$= [(2j + 1)(2J' + 1)(2J'' + 1)(2J + 1)]^{\frac{1}{2}} W(\lambda J''Jj; J'J)$$

$$\times W(\lambda j J'j; jJ'') \quad (12.20)$$

From the antisymmetry of the total wave function it is clear that each of the operators $\Omega(2)$ and $\Omega(3)$ make exactly the same contribution as that given in (12.20).[5] Consequently,

$$\langle \Omega \rangle = 3C(J\lambda J; M0) \sum_{J'J''} F_{J'}F_{J''}(J \parallel \Omega(1) \parallel J) \quad (12.21)$$

The interesting quantity is the ratio

$$\mathfrak{R} = \langle \Omega \rangle_{M=J}/(j \parallel \Omega(1) \parallel j) \, C(j\lambda j; j0) \quad (12.22)$$

which is the ratio of the matrix element for the three-particle configuration to its value for a single particle. In Table 2 we give the results for $\mathfrak{R}$ for the configurations discussed above and for $\lambda = 1$ and 2.

Table 2.   Ratio of Matrix Elements for Three- and One-Particle Configurations

Configuration	J	$\lambda = 1$	$\lambda = 2$
$(\frac{3}{2})^3$	$\frac{3}{2}$	1	-1
$(\frac{5}{2})^3$	$\frac{3}{2}$	$\frac{3}{5}$	0
	$\frac{5}{2}$	1	0
	$\frac{9}{2}$	$\frac{9}{5}$	0

For $j = J$ a somewhat simpler result applies. Then

$$\mathfrak{R} = 3(2J + 1) \sum_{J'J''} F_{J'}F_{J''}[(2J' + 1)(2J'' + 1)]^{\frac{1}{2}} W^2(\lambda J'JJ; J''J)$$

For $\lambda = 1$ the conditions that $|J' - J''|$ is an even integer and that $\Delta(1J'J'')$ exists imply $J' = J''$. Then, from Table I.4,

$$W^2(1J'JJ; J'J) = \frac{J'(J' + 1)}{4J(J + 1)(2J + 1)(2J' + 1)}$$

---

[5] As a check we see that for $\Omega(2)$ everything in (12.19) would be unchanged except that the last C-coefficient would be replaced by $C(j\lambda j; m-m',0)$. If we now set $m - m' = \nu'$, so that ν' replaces m' as a summation index, and apply (3.16b) to the second and fourth C-coefficients, we find the sum in (12.19) reproduced with the multiplicative phase factor $(-)^{2j-J'}(-)^{2j-J''} = (-)^{J'-J''} = 1$, since J' and J'' are both even integers.

and we find

$$\mathcal{R} = \frac{3}{4J(J+1)} \sum_{J'} J'(J'+1)F^2_{J'}. = 1$$

For $\lambda = 2$ this simple result does not apply but we may note that $|J' - J''| = 0, 2$ only.

By using the properties of the Racah coefficients described in section 23 it is possible to reduce the result (12.20) to the form

$$(J \parallel \Omega(1) \parallel J)/(j \parallel \Omega(1) \parallel j)$$

$$= [(2j+1)(2J+1)]^{\frac{1}{2}} W(J'Jj\lambda; jJ)\, \delta_{J'J''} \quad (12.20')$$

This can also be obtained from direct evaluation of the matrix element of $\Omega(3)$. Then for first-rank tensors we may deduce, with the aid of Table I.4, that for configurations of three equivalent particles

$$\mathcal{R} = J/j$$

The results for $\lambda = 1$ apply to the magnetic dipole moment, and $\mathcal{R}$ is simply J/j. It is clear that in the exceptional case $j \neq J$ we can attribute considerable departure from the Schmidt limits (corresponding to a single odd particle) to the model considered here. This has been pointed out before.[6] It should not be assumed, however, that other effects (configuration mixing, for example)[7] are not equally important, and the results implied by the foregoing discussion are not necessarily those that compare best with observations.

The results for $\lambda = 2$ apply to the quadrupole moment. The reversal of sign for the $(\frac{3}{2})^3$ configuration is to be understood quite simply since this configuration is a single hole in the $\frac{3}{2}$ shell. This sign reversal is seen to be characteristic of second-rank tensors. The vanishing of the matrix elements of a second-rank tensor for the $(\frac{5}{2})^3$ configuration is due to the fact that this is a half-filled shell. Some of these results have appeared elsewhere in the literature.[8] It should be realized that for the quadrupole moment contributions of the distorted core are very essential. The discussion presented in this section is meant to be illustrative

[6] See M. G. Mayer and J. H. D. Jensen, *Elementary Theory of Nuclear Shell Structure*, John Wiley & Sons, New York, 1955. As these authors show the simple result $\mathcal{R} = J/j$ can be obtained directly by taking the sum of single particle expectation values and noting that the reduced matrix element (essentially the gyromagnetic ratio in the case of the magnetic moment) is the same for all nucleons in equivalent orbits.

[7] R. J. Blin-Stoyle and M. A. Perks, *Proc. Phys. Soc.* **A 67**, 885 (1954).

[8] A. Bohr and B. R. Mottelson, *Kgl. Danske Videnskab. Selskab, Math.-fys. Medd.* **27**, no. 16 (1953).

rather than a thoroughgoing investigation of all pertinent physical factors affecting the calculation of the moments.

We can also study transitions between states $J_1 M_1 \rightarrow J_2 M_2$ where $J_1 \neq J_2$ necessarily. For example, in gamma transitions the initial and/or final configuration may be one of equivalent particles. As was discussed in section 27, the matrix elements for emission of 2^L pole radiation involves a tensor of rank L. In the case of three equivalent particles, if we remember that j must be the same for the initial and final states, since the interaction is a sum of one-particle operators, the appropriate matrix element is

$$(J_2 M_2 | \Omega_L | J_1 M_1) = C(J_1 L J_2; M_1 - M)(J_2 \| \Omega_L \| J_1)$$

where $M = M_1 - M_2$ refers to the radiation. The reduced matrix element is given by

$$(J_2 \| \Omega_L \| J_1)/(j \| \Omega_L \| j)$$

$$= 3 \sum_{J'J} F_{J'}(J_1) \, F_{J''}(J_2)[(2j + 1)(2J' + 1)(2J'' + 1)(2J_1 + 1)]^{\frac{1}{2}}$$

$$\times W(\lambda J'' J_1 j; J' J_2) \, W(\lambda j J j; j J'') \quad (12.20a)$$

which reduces to (12.20) when $J_1 = J_2 = J$. We have indicated that the fractional parentage coefficients in initial and final state may depend on the resultant angular momentum.

The transition probability is proportional to

$$w = \frac{1}{2J_1 + 1} \sum_{M_1 M_2} |(J_2 M_2 | \Omega_L | J_1 M_1)|^2$$

$$= \frac{2J_2 + 1}{2J_1 + 1} (J_2 \| \Omega_L \| J_1)^2 \quad (12.20b)$$

Thus, the square of (12.20a) gives the factor by which the single-particle transition probability must be multiplied to give the value appropriate to the three-particle configurations with equivalent orbits.

38. IDENTICAL PARTICLES IN L–S COUPLING

The discussion of the preceding section can readily be extended to the treatment of the coupling of three equivalent identical particles in L–S coupling.[9] We denote the orbital and intrinsic angular momenta by $s_1 l_1$, $s_2 l_2$, and $s_3 l_3$.

[9] The standard reference on this subject is Condon and Shortley, *Theory of Atomic Spectra*, Cambridge University Press, 1935. See also G. Racah, *Phys. Rev.* **61**, 186 (1942); **62**, 438 (1942); **63**, 367 (1943).

If we leave aside, for the moment, all questions of symmetrization, the unitary transformation between the schemes

$$s_1 + s_2 = S'$$

$$l_1 + l_2 = L'$$

$$S' + s_3 = S$$

$$L' + l_3 = L \qquad (12.23a)$$

and

$$s_2 + s_3 = S''$$

$$l_2 + l_3 = L''$$

$$s_1 + S'' = S$$

$$l_1 + L'' = L \qquad (12.23b)$$

is the product $R_{S''S'}R_{L''L'}$ where, see equation (6.3)

$$R_{S''S'} = [(2S'' + 1)(2S' + 1)]^{\frac{1}{2}} W(s_1 s_2 S s_3; S'S'')$$

$$R_{L''L'} = [(2L'' + 1)(2L' + 1)]^{\frac{1}{2}} W(l_1 l_2 L l_3; L'L'')$$

This follows from the fact that S and L are decoupled. Thus, for the two schemes, (12.23a) and (12.23b), we may write

$$\Psi_{SLM}(s_1 l_1, s_2 l_2 (S'L') s_3 l_3; SLM) \equiv \Psi_{SLM}(S'L')$$

$$= \sum_{S''L''} R_{S''S'} \, R_{L''L'} \, \Psi_{SLM}(S''L'') \quad (12.24)$$

where

$$\Psi_{SLM}(S''L'') \equiv \Psi(s_1 l_1, s_2 l_2, s_3 l_3 (S''L''); SLM)$$

For definiteness we assume that we deal with fermions. Taking the antisymmetrization condition into account, and assuming equivalent particles, $s_i = s$, $l_i = l$, it is clear that interchange of particles 2 and 3, for example, multiplies $\Psi_{SLM}(S''L'')$ by $(-)^{2s-S''}(-)^{2l-L''}$—see (12.4). In order that this factor be equal to -1 for $s = \frac{1}{2}$ and l integer, it is necessary that

$$(-)^{S''+L''} = 1$$

or $S'' + L''$ is an even integer. This leads, for example, to allowed states ^1S, ^3P, and ^1D for the p^2 configuration. To construct a wave function antisymmetric in all particles, we proceed as in the case of j–j

coupling. Such a wave function will be a linear combination of $\Psi_{SLM}(S'L')$ for various $S'L'$. Thus,

$$\Psi_{SLM} = \sum_{S'L'} F_{S'L'} \Psi_{SLM}(S'L')$$

$$= \sum_{S''L''} \sum_{S'L'} R_{S''S'} R_{L''L'} F_{S'L'} \Psi_{SLM}(S''L'') \qquad (12.25)$$

Since the coefficient of each $\Psi_{SLM}(S''L'')$ for non-allowed $S''L''$-combinations must vanish, we require that

$$\sum_{S'L'} R_{S''S'} R_{L''L'} F_{S'L'} = 0 \quad \text{for} \quad S'' + L'' \text{ odd} \qquad (12.26)$$

Of course $S' + L'$ is also even, so that antisymmetry relative to both operators P_{13} and P_{23} is assured.

As an example, consider the configuration p^3 so that $l_1 = l_2 = l_3 = 1$, $s_1 = s_2 = s_3 = \frac{1}{2}$. We construct the antisymmetric wave function for the ^2P state ($S = \frac{1}{2}$, $L = 1$). The possible values of $S''L''$ with both even and odd sum are given in Table 3. We write

$$\Psi_{\frac{1}{2}1}(11) = \sum_{n=1}^{6} a_n \psi_n$$

where $\psi_1 \cdots \psi_6$ are $\Psi_{\frac{1}{2}1}(S''L'')$ for the six values of $S''L''$ listed in the order of Table 3. Similarly,

$$\Psi_{\frac{1}{2}1}(02) = \sum_{n=1}^{6} b_n \psi_n$$

$$\Psi_{\frac{1}{2}1}(00) = \sum_{n=1}^{6} c_n \psi_n$$

Table 3. $S''L''$ Values for the p^2 Configuration*

Term Designation	S''	L''	Wave Function
^3D	1	2	ψ_1
^1P	0	1	ψ_2
^3S	1	0	ψ_3
^1D	0	2	ψ_4
^3P	1	1	ψ_5
^1S	0	0	ψ_6

* The last three $S''L''$ entries in the table give the permissible states for which the sum is even and are possible values of S' and L'.

Then we find by (12.26) the results given in Table 4.

Table 4.   Coefficients in L–S Coupling

n	a_n	b_n	c_n
1	$\frac{1}{4}\sqrt{\frac{5}{3}}$	$\frac{1}{4}\sqrt{\frac{1}{3}}$	$\frac{1}{2}\sqrt{\frac{5}{3}}$
2	$\frac{1}{4}\sqrt{3}$	$-\frac{1}{3}\sqrt{\frac{5}{3}}$	$\frac{1}{2}\sqrt{\frac{1}{3}}$
3	$-\frac{1}{2}\sqrt{\frac{1}{3}}$	$\frac{1}{2}\sqrt{\frac{5}{3}}$	$\frac{1}{2}\sqrt{\frac{1}{3}}$
4	$\frac{1}{4}\sqrt{5}$	$-\frac{1}{12}$	$-\frac{1}{6}\sqrt{5}$
5	$\frac{1}{4}$	$\frac{1}{4}\sqrt{5}$	$-\frac{1}{2}$
6	$-\frac{1}{2}$	$-\frac{1}{6}\sqrt{5}$	$-\frac{1}{6}$

The totally antisymmetrical wave function is

$$\Psi(^2P) = F_{11}\,\Psi_{\frac{3}{2}1}(11) + F_{02}\,\Psi_{\frac{3}{2}1}(02) + F_{00}\,\Psi_{\frac{3}{2}1}(00) \qquad (12.27)$$

In this expression the coefficients of ψ_1, ψ_2, and ψ_3 must vanish.   Hence

$$F_{11}a_1 + F_{02}b_1 + F_{00}c_1 = 0$$

$$F_{11}a_2 + F_{02}b_2 + F_{00}c_2 = 0 \qquad (12.28)$$

$$F_{11}a_3 + F_{02}b_3 + F_{00}c_3 = 0$$

It may be verified that the determinant of the coefficients in (12.28) does indeed vanish.   Then with the normalization condition

$$F_{11}^2 + F_{02}^2 + F_{00}^2 = 1$$

we find the fractional parentage coefficients

$$F_{11} = \sqrt{\tfrac{1}{2}} \qquad F_{02} = \tfrac{1}{3}\sqrt{\tfrac{5}{2}}, \qquad F_{00} = -\tfrac{1}{3}\sqrt{2} \qquad (12.29)$$

where we arbitrarily choose $F_{11} > 0$.[10]

Wave functions for other configurations can be constructed in a similar fashion.[10]   The calculation of various moments in L–S coupling, as was done above for j–j coupling, is then readily carried out.

It will be seen that the procedure for determining the fractional parentage coefficients in L–S coupling is not essentially different from that used for j–j coupling.

[10] The permissible terms for p^n-, d^n-, and f^n-configurations [$1 \leq n \leq 2(2l + 1)$] are given, for example, in H. E. White, *Introduction to Atomic Spectra*, p. 437 McGraw-Hill Book Co., New York, 1934. For p^3 these are ^2P, ^2D, and ^4S. The other terms which are consistent with vector addition rules lead to inconsistent equations wherein the determinant of the coefficients does not vanish.

For further discussion and procedures applying to more complicated configurations the basic papers of Racah [9] should be consulted.[11]

39. THE ISOTOPIC SPIN

The concept of the isotopic spin has often been used in nuclear physics to express the symmetry that would pertain if nuclear forces were independent of the charge state of the nucleons.[12] Then for light nuclei, for which, in first approximation, we might neglect the electrostatic Coulomb repulsion between protons, all the nucleons are essentially the same kind of particle. The distinction between a neutron and proton then lies in the fact that they represent different substates in charge space. That is, each nucleon is characterized by a charge wave function $\varphi_{i\nu}$ where ν takes on the values

$$-i \leq \nu \leq i$$

Since there are only two charge states, we see that $2i + 1 = 2$ and the isotopic spin $i = \frac{1}{2}$. We can arbitrarily assign $\nu = \frac{1}{2}$ to the neutron and $\nu = -\frac{1}{2}$ to the proton. Then in detailed form the charge wave functions would be

$$\varphi_{\frac{1}{2}\frac{1}{2}} = \begin{pmatrix} 1 \\ 0 \end{pmatrix} \qquad \varphi_{\frac{1}{2}-\frac{1}{2}} = \begin{pmatrix} 0 \\ 1 \end{pmatrix} \tag{12.30}$$

These correspond to neutron and proton states, respectively.

The formal equivalence to the theory of (non-relativistic) spin $\frac{1}{2}$ particles is obvious. Thus, the wave functions (12.30) diagonalize the z-component of isotopic spin, $i_z = \frac{1}{2}\tau_z$ where

$$\tau_z = \begin{pmatrix} 1 & 0 \\ 0 & -1 \end{pmatrix} \tag{12.31a}$$

Also, τ_x and τ_y have the same form as the Pauli spin operators σ_x and σ_y

$$\tau_x = \begin{pmatrix} 0 & 1 \\ 1 & 0 \end{pmatrix}, \qquad \tau_y = \begin{pmatrix} 0 & -i \\ i & 0 \end{pmatrix} \tag{12.31b}$$

in the representation which diagonalizes τ_z. The commutation rules are $\boldsymbol{\tau} \times \boldsymbol{\tau} = 2i\boldsymbol{\tau}$ just as for the Pauli $\boldsymbol{\sigma}$-operators. Then

$$\tau_+ = \frac{1}{2}(\tau_x + i\tau_y) = \begin{pmatrix} 0 & 1 \\ 0 & 0 \end{pmatrix} \tag{12.32a}$$

[11] Also Condon and Shortley, reference 9 and S. Meshkov, *Phys. Rev.* **91**, 871 (1953).

[12] One of the earlier discussions was given by B. Cassen and E. U. Condon, *Phys. Rev.* **50**, 846 (1936); see also any standard reference in nuclear physics, for example, L. Rosenfeld, *Nuclear Forces*, Vol. I, North Holland Publishing Co.

and

$$\tau_- = \tfrac{1}{2}(\tau_x - i\tau_y) = \begin{pmatrix} 0 & 0 \\ 1 & 0 \end{pmatrix} \tag{12.32b}$$

have the following properties:

$$\tau_+\varphi_{\frac{1}{2}\frac{1}{2}} = 0, \qquad\qquad \tau_+\varphi_{\frac{1}{2}-\frac{1}{2}} = \varphi_{\frac{1}{2}\frac{1}{2}}$$

$$\tau_-\varphi_{\frac{1}{2}\frac{1}{2}} = \varphi_{\frac{1}{2}-\frac{1}{2}}, \qquad \tau_-\varphi_{\frac{1}{2}-\frac{1}{2}} = 0 \tag{12.33}$$

or

$$\tau_{\mp}\varphi_{\frac{1}{2}\nu} = (\tfrac{1}{2} \pm \nu)\,\varphi_{\frac{1}{2}-\nu} = (-)^{\nu-\frac{1}{2}}(\tfrac{3}{2})^{\frac{1}{2}}\,C(\tfrac{1}{2}1\tfrac{1}{2}; \nu, \mp 1)\,\varphi_{\frac{1}{2}-\nu}$$

Thus τ_+ creates a neutron when applied to a proton and gives zero when applied to a neutron; τ_- has the reverse property of transforming a neutron to a proton. The role of these operators in beta decay will be apparent; see Chapter IX.

We now consider configurations of two particles. Then [13]

$$N = \sum_i \nu_i$$

where the sum is over both particles and has the possible values ± 1 and 0. From analogy with the ordinary spin, one expects the total isotopic spin I, where $I(I + 1)$ is the eigenvalue of $\mathbf{I}^2$ and $\mathbf{I} = \mathbf{i}_1 + \mathbf{i}_2$, to have the values 0 and 1. The total charge wave function is then

$$\Phi_{IN} = \sum_\nu C(\tfrac{1}{2}\tfrac{1}{2}I; \nu, N-\nu)\,\varphi_{i\nu}(1)\,\varphi_{i\,N-\nu}(2) \tag{12.34}$$

The isotopic spin triplet $I = 1$ has the detailed form:

$$\Phi_{11} = \varphi_{\frac{1}{2}\frac{1}{2}}(1)\,\varphi_{\frac{1}{2}\frac{1}{2}}(2) \qquad\qquad\qquad \text{2 neutrons}$$

$$\Phi_{10} = \frac{1}{\sqrt{2}}\,[\varphi_{\frac{1}{2}\frac{1}{2}}(1)\,\varphi_{\frac{1}{2}-\frac{1}{2}}(2) + \varphi_{\frac{1}{2}-\frac{1}{2}}(1)\,\varphi_{\frac{1}{2}\frac{1}{2}}(2)] \quad \text{neutron} + \text{proton}$$

$$\Phi_{1-1} = \varphi_{\frac{1}{2}-\frac{1}{2}}(1)\,\varphi_{\frac{1}{2}-\frac{1}{2}}(2) \qquad\qquad\qquad \text{2 protons} \tag{12.35}$$

The isotopic spin singlet wave function is

$$\Phi_{00} = \frac{1}{\sqrt{2}}\,[\varphi_{\frac{1}{2}\frac{1}{2}}(1)\,\varphi_{\frac{1}{2}-\frac{1}{2}}(2) - \varphi_{\frac{1}{2}-\frac{1}{2}}(1)\,\varphi_{\frac{1}{2}\frac{1}{2}}(2)] \quad \text{neutron} + \text{proton}$$

If the two nucleons constitute the entire nucleus, the first and third correspond to ^1S states of the unbound systems, and Φ_{10} is the ^1S state

[13] In general, for Z protons and $A-Z$ neutrons, $N = \tfrac{1}{2}(A - 2Z)$. Since this is one-half the difference of neutron number and proton number, it would be more appropriate to refer to isobaric spin rather than isotopic spin.

of the deuteron; Φ_{00}, the isotopic spin singlet, is the $(^3S + {}^3D)$ ground state of the deuteron. Since interchange of charge coordinates gives

$$P_{12}^c \Phi_{IN} = (-)^{I+1} \Phi_{IN}$$

the triplet charge state is symmetric and the singlet antisymmetric with respect to charge coordinate exchange. This, of course, is obvious from (12.35). But, on the above assignments, the total wave function, which is a product of the charge wave function Φ_{IN} and the spin-space wave function, is seen to be antisymmetric in every case, where interchange of all coordinates is implied. This is what is to be expected since all nucleons are now indistinguishable.

As for the form of the actual antisymmetrical wave functions, we consider the case of two j–j coupled nucleons. If these are non-equivalent nucleons, in the sense of section 37, the space wave function is

$$\Psi_{JM} = \frac{1}{\sqrt{2}} \sum_m C(j_1 j_2 J; m, M - m)$$

$$\times [\psi_{j_1 m}(1)\, \psi_{j_2\, M-m}(2) \pm \psi_{j_1 m}(2)\, \psi_{j_2\, M-m}(1)] \quad (12.36)$$

according to whether symmetry $(+)$ or antisymmetry $(-)$ in spin-coordinate space is required. Since the total wave function is $\Psi = \Phi_{IN}\Psi_{JM}$, it is clear that the $+$ sign in (12.36) is taken with $I = 0$ and the $-$ sign with $I = 1$. Hence

$$\Psi = \frac{1}{\sqrt{2}} \Phi_{IN} \sum_m C(j_1 j_2 J; m, M - m)$$

$$\times [\psi_{j_1 m}(1)\, \psi_{j_2\, M-m}(2) + (-)^I \psi_{j_1 m}(2)\, \psi_{j_2\, M-m}(1)] \quad (12.37)$$

If the two nucleons are in equivalent orbits, this result can be written more simply in the form

$$\Psi = \Phi_{IN} \sum_m C(jjJ; m, M - m)\, \psi_{jm}(1)\, \psi_{j\, M-m}(2) \quad (12.38)$$

and, since

$$P_{12}\, \Psi = (-)^{I+1}(-)^{2j-J}\, \Psi$$

and $2j$ is an odd integer, we have overall antisymmetry if $J + I$ is odd.

As an application of these results consider a beta transition from an initial state with quantum numbers j_1', j_2', J', I' to a final state with quantum numbers j_1, j_2, J, and I. The matrix element is to be calculated for a transition between these states for a beta operator

$$H_\beta = \sum_i \Omega_{\lambda M}(i)\, \tau_-(i), \qquad i = 1, 2 \quad (12.39)$$

where λ is the rank of the tensor and i labels the nucleons. For definiteness we consider β^- emission and, hence, use τ_- in (12.39).

For simplicity assume that $j_1'' = j_2'' = j'$ but that $j_1 \neq j_2$ necessarily. Then, following a procedure exactly like that used in section 37 leading to (12.21), we find

$$(Jm \,|\, H_\beta \,|\, J'm') = C(J'\lambda J; m'M)(J \parallel H_\beta \parallel J') \qquad (12.40)$$

and the reduced matrix element is [14]

$$(J \parallel H_\beta \parallel J') = (-)^{\lambda-J}(2J' + 1)^{\frac{1}{2}}[(2j_1 + 1)^{\frac{1}{2}} \, \delta_{j_2 j'} \, W(j'j_1 J'J; \lambda j')$$

$$\times (j_1 \parallel \Omega_\lambda \parallel j') + (2j_2 + 1)^{\frac{1}{2}} \, \delta_{j_1 j'}(-)^{j_2+j'+J+I}$$

$$\times W(j'j_2 J'J; \lambda j')(j_2 \parallel \Omega_\lambda \parallel j')] \qquad (12.40')$$

The Kronecker delta symbols in (12.40') mean that all the individual angular momentum quantum numbers (j and l, for example) must be alike in initial and final states for at least one of the nucleons. Since we assumed $j_1 \neq j_2$, it follows that one or the other of the two terms in (12.40') must vanish. In that case, apart from a rather unessential phase, the result has the same form, no matter which particle changes its quantum numbers.

If for both the initial and final states the particles are in equivalent orbits, the result is

$$(J \parallel H_\beta \parallel J') = (-)^{\lambda-J}[(2J' + 1)(2j + 1)]^{\frac{1}{2}}$$

$$\times W(j'jJ'J; \lambda j')(j \parallel H_\beta \parallel j') \, \delta_{jj'} \qquad (12.41)$$

where $j_1 = j_2 = j$.

The interesting feature of these results is the distinction between Fermi and Gamow–Teller interactions for allowed transitions.[15] In that case $\lambda = 0$ for Fermi interactions ($\Omega = \beta \approx -1$ or $\Omega = 1$) and $\lambda = 1$ for Gamow–Teller interactions ($\Omega_{1M} = \beta\sigma_M \approx -\sigma_M$ or $\Omega_{1M} = \sigma_M$). For $\lambda = 0$ we have $J' = J$ as expected. But also $j_1 = j'$ in the first term of (12.40'), and $j_2 = j'$ in the second term. Consequently, since there can be no parity change in the transition considered, it follows that equivalent orbits in the initial (final) state implies equivalent orbits in the final (initial) state. Since both $I + J$ and $I' + J'$ are odd and $J = J'$, it follows that I and I' are both even or both odd. But, since $|\Delta I| = 0, 1$ it follows finally that

$$\Delta I = 0$$

[14] M. E. Rose and R. Osborne, *Phys. Rev.* **93**, 1326 (1954).

[15] Cf. for example, E. J. Konopinski, *Beta and Gamma Ray Spectroscopy*, Chapter X, North Holland Publishing Co., 1955.

for Fermi interactions with equivalent orbits. The $0 \to 0$ transition in C^{10}, O^{14}, and Cl^{34} are good examples of this selection rule. These transitions occur between corresponding levels of two members of the triads C^{10}—B^{10}—Be^{10}, O^{14}—N^{14}—C^{14}, A^{34}—Cl^{34}—S^{34}. These levels clearly belong to $I = 1$, and therefore $\Delta I = 0$.

In the Gamow–Teller case ($\lambda = 1$), we see that $|j' - j_1| = 0, 1$ when $j_2 = j'$—cf. equation (12.40'). The fact that the parity does not change constitutes no restriction on $|j' - j_1|$. However, even if $j' = j_1$, we can still have $|J - J'| = 0, 1$ and $|\Delta I| = 0, 1$. However, there is no doubt that for the light nuclei of mass number $4n+2$ the pure Gamow–Teller transitions ($\Delta J = \pm 1$) correspond to $\Delta I = \pm 1$. In the transition $He^6 \to Li^6$, where we expect the two nucleons outside the alpha-particle core to be $p_{\frac{3}{2}}$-particles, we have $J + I$ and $J' + I'$ odd. Since $\Delta J = 1$, it follows that $\Delta I = -1$. In fact, the He^6 ground state is a member of a triad, the other two substates corresponding to $N = 0$ and -1 being the first excited state of Li^6 and the ground state of Be^6. So $I' = 1$. The ground state of Li^6 is an isotopic spin singlet and $I = 0$, in accord with the foregoing. The C^{14} decay, to the ground state of N^{14} is exactly the same sort of situation, and the same remark applies to $O^{14} \to N^{14}$ (ground).

For the transitions between mirror nuclei in the p-shell, it is clear that $\Delta J = 0$ and $\Delta I = 0$. These all have odd mass and half-integer J so that both Fermi and Gamow–Teller interactions contribute. For more complex configurations no simple isotopic spin selection rules appear possible.

APPENDIX I.   CLEBSCH–GORDAN
AND RACAH COEFFICIENTS

A. *C*-COEFFICIENTS

Since the Clebsch–Gordan or C-coefficients are basic for the theory of angular momentum, we wish to extend the discussion of their explicit determination. The derivation of the symmetry rules for these coefficients was given in section 11 and was based on the result (3.19) given by Racah.[1] Although no attempt is made to reproduce all steps of Racah's analysis, we shall reproduce his procedure in outline and discuss the phase question which arises in defining the C-coefficients. It will be recalled that the procedure of section 12 also defines the C-coefficients once the phase question is settled.

First we obtain a recurrence formula across a row—this means for fixed j_1, j_2, and j by applying $J_\pm$ to (3.6). Here, of course, $\mathbf{J} = \mathbf{J}_1 + \mathbf{J}_2$.

$$[(j \mp m)(j \pm m + 1)]^{\frac{1}{2}} C(j_1 j_2 j; m_1, m \pm 1 - m_1)$$

$$= [(j_1 \mp m_1 + 1)(j_1 \pm m_1)]^{\frac{1}{2}} C(j_1 j_2 j; m_1 \mp 1, m - m_1 \pm 1)$$

$$+ [(j_2 \mp m \pm m_1)(j_2 \pm m \mp m_1 + 1)]^{\frac{1}{2}} C(j_1 j_2 j; m_1, m - m_1) \quad (I.1)$$

We follow Racah's [1] procedure and define a quantity f by

$$C(j_1 j_2 j; m_1 m_2) = (-)^{j_1 - m_1} f[(j_1 + m_1)!(j_2 + m_2)!(j + m)!]^{\frac{1}{2}}$$

$$\times [(j_1 - m_1)!(j_2 - m_2)!(j - m)!]^{-\frac{1}{2}} \quad (I.2)$$

and $m_1 + m_2 = m$. After introducing this into (I.1), we find by setting $m = j$ in the equation with the upper signs that f for $m = j$ is independent of m_1 and m_2. Using the result derived from the lower signs in (I.1), we obtain a simple recurrence formula for f, the solution of which is

$$f = f_j \frac{(j_1 - m_1)!(j_2 - m_2)!}{(j_1 + m_1)!(j_2 + m_2)!}$$

$$\times \sum_t (-)^t \binom{j - m}{t} \frac{(j_1 + m_1 + t)!(j_2 + m_2 + j - m - t)!}{(j_1 - m_1 - t)!(j_2 - m_2 - j + m + t)!} \quad (I.3)$$

[1] G. Racah, *Phys. Rev.* **62**, 438 (1942).

The determination of $f_j \equiv f(m = j)$ depends on the use of a recurrence formula in j. We outline the method (due to Condon and Shortley [2]) sufficiently to permit the reader to reconstruct the details.

First we apply J_{1z} to ψ_{jm}. This gives

$$J_{1z} \psi_{jm} = \sum_{j'm'} (j'm'|J_{1z}|jm) \, \psi_{j'm'}$$

$$= \sum_{\mu} \sum_{j'm'} (j'm'|J_{1z}|jm) \, C(j_1 j_2 j'; \mu, m' - \mu) \, \psi_{j_1\mu} \psi_{j_2 \, m' - \mu}$$

$$= \sum_{\mu} \mu \, C(j_1 j_2 j; \mu, m - \mu) \, \psi_{j_1\mu} \psi_{j_2 \, m - \mu}$$

Whence

$$m_1 \, C(j_1 j_2 j; m_1, m - m_1) = \sum_{j'} (j'm'|J_{1z}|jm) \, C(j_1 j_2 j'; m_1, m - m_1) \quad \text{(I.4)}$$

The sum in (I.4) goes through the values $j' = j, \, j \pm 1$ as is obvious from the fact that J_{1z} is a first-rank tensor component; see Chapter V and section 19. For $j' = j$ a simple evaluation of the matrix element is possible. We use (5.29) to write

$$j(j + 1)(jm|J_{1z}|jm) = (jm|J_z|jm)(j \, \| \, \mathbf{J_1 \cdot J} \, \| \, j)$$

The first factor on the right is simply m. The second factor is evaluated from squaring $\mathbf{J_2} = \mathbf{J} - \mathbf{J_1}$. Hence

$$(j \, \| \, \mathbf{J_1 \cdot J} \, \| \, j) = \tfrac{1}{2}[j(j + 1) + j_1(j_1 + 1) - j_2(j_2 + 1)]$$

since the reduced matrix element of a zero-rank tensor (which is necessarily diagonal) is equal to its eigenvalue.

To obtain the matrix elements of J_{1z} for $j = j' \pm 1$, we first recall that

$$(j'm'|J_{1z}|jm) = C(j1j'; m0) \, \delta_{mm'} (j \, \| \, J_1 \, \| \, j') \quad \text{(I.5)}$$

We obtain two relations between the two unknown (reduced) matrix elements $(j \, \| \, J_1 \, \| \, j \pm 1)$ as follows. First the matrix element of

$$[J_{1-}, J_{1z}] = J_{1-} \quad \text{(I.6)}$$

is taken between the states ψ_{jm} and $\psi_{j \, m+1}$. Since there is a product of two $\mathbf{J}$-operators on the left, this will be a quadratic equation in the unknown. It has the form

$$\sum_{j'm'} [(jm|J_{1-}|j'm')(j'm'|J_{1z}|jm + 1)$$

$$- (jm|J_{1z}|j'm')(j'm'|J_{1-}|jm + 1)] = (jm|J_{1-}|jm + 1) \quad \text{(I.7)}$$

[2] E. U. Condon and G. H. Shortley, *Theory of Atomic Spectra*, Cambridge University Press, 1935.

In the first term on the left, $m' = m + 1$, and in the second $m' = m$. A second equation is obtained by taking the diagonal matrix element of

$$\mathbf{J}_1^2 = J_{1z}^2 + \tfrac{1}{2}(J_{1+}J_{1-} + J_{1-}J_{1+})$$

The calculation of this matrix element is somewhat different from the simpler one carried out in section 12. It is clear that there will again be a sum over $j' = j, j \pm 1$ and a trivial sum over m', and that the reduced matrix elements for $j' \neq j$ enter to second degree. The ambiguity of sign is resolved by choosing $(j \parallel J_1 \parallel j - 1) = -(j \parallel J_2 \parallel j - 1)$ real and positive. The reduced matrix element $(j \parallel J_1 \parallel j + 1) = (j + 1 \parallel J_1 \parallel j)$ is obtained from this by replacing j by $j + 1$. The recurrence formula in j that results is

$$\left[m_1 - m\frac{j_1(j_1 + 1) - j_2(j_2 + 1) + j(j + 1)}{2j(j + 1)} \right] C(j_1 j_2 j; m_1, m - m_1)$$

$$= \left[\frac{\left\{ \begin{matrix} (j^2 - m^2)(j - j_1 + j_2)(j + j_1 - j_2) \\ \times\, (j_1 + j_2 + j + 1)(j_1 + j_2 - j + 1) \end{matrix} \right\}}{4j^2(2j - 1)(2j + 1)} \right]^{\frac{1}{2}} C(j_1 j_2 j - 1; m_1, m - m)$$

$$+ \left[\frac{\left\{ \begin{matrix} [(j + 1)^2 - m^2](j + 1 - j_1 + j_2)(j + 1 + j_1 - j_2) \\ \times\, (j_1 + j_2 + j + 2)(j_1 + j_2 - j) \end{matrix} \right\}}{4(j + 1)^2(2j + 1)(2j + 3)} \right]^{\frac{1}{2}}$$

$$\times\, C(j_1 j_2 j + 1; m_1, m - m_1) \quad \text{(I.8)}$$

With the results obtained by essentially this procedure Racah was able to establish a recurrence formula for f_j whose solution is determined by the convention (3.11a). Further transformations to obtain the result (3.19) are described in references 1 and 3.

Since explicit results for $C(j_1 j_2 j; m_1, m - m_1)$ for $j_2 = \tfrac{1}{2}$ and 1 are very often quite useful, these are given below in Tables I.1 and I.2.

Table I.1. $C(j_1 \tfrac{1}{2} j; m - m_2, m_2)$

$j =$	$m_2 = \tfrac{1}{2}$	$m_2 = -\tfrac{1}{2}$
$j_1 + \tfrac{1}{2}$	$\left[\dfrac{j_1 + m + \tfrac{1}{2}}{2j_1 + 1}\right]^{\frac{1}{2}}$	$\left[\dfrac{j_1 - m + \tfrac{1}{2}}{2j_1 + 1}\right]^{\frac{1}{2}}$
$j_1 - \tfrac{1}{2}$	$-\left[\dfrac{j_1 - m + \tfrac{1}{2}}{2j_1 + 1}\right]^{\frac{1}{2}}$	$\left[\dfrac{j_1 + m + \tfrac{1}{2}}{2j_1 + 1}\right]^{\frac{1}{2}}$

3 G. Racah, *Phys. Rev.* **61**, 186 (1942).

Table I.2. $C(j_1 1j; m-m_2, m)$

$j =$	$m_2 = 1$	$m_2 = 0$	$m_2 = -1$
$j_1 + 1$	$\left[\dfrac{(j_1+m)(j_1+m+1)}{(2j_1+1)(2j_1+2)}\right]^{\frac{1}{2}}$	$\left[\dfrac{(j_1-m+1)(j_1+m+1)}{(2j_1+1)(j_1+1)}\right]^{\frac{1}{2}}$	$\left[\dfrac{(j_1-m)(j_1-m+1)}{(2j_1+1)(2j_1+2)}\right]^{\frac{1}{2}}$
j_1	$-\left[\dfrac{(j_1+m)(j_1-m+1)}{2j_1(j_1+1)}\right]^{\frac{1}{2}}$	$\dfrac{m}{[j_1(j_1+1)]^{\frac{1}{2}}}$	$\left[\dfrac{(j_1-m)(j_1+m+1)}{2j_1(j_1+1)}\right]^{\frac{1}{2}}$
$j_1 - 1$	$\left[\dfrac{(j_1-m)(j_1-m+1)}{2j_1(2j_1+1)}\right]^{\frac{1}{2}}$	$-\left[\dfrac{(j_1-m)(j_1+m)}{j_1(2j_1+1)}\right]^{\frac{1}{2}}$	$\left[\dfrac{(j_1+m+1)(j_1+m)}{2j_1(2j_1+1)}\right]^{\frac{1}{2}}$

B. RACAH COEFFICIENTS

Here we summarize some useful properties of the Racah coefficients.

B1. Relation to Legendre Polynomials

In angular distributions wherein radiation of angular momentum L is emitted in a transition from a state with angular momentum j' to a state with angular momentum j, the following Racah coefficient enters (see section 33):

$$W(jjLL; \nu j') = (-)^{j+L-\nu-j'} W(L\nu j'j; Lj)$$

see B2 below.   Since

$$\mathbf{j'} = \mathbf{L} + \mathbf{j}$$

we can define the angle θ between $\mathbf{L}$ and $\mathbf{j}$ by

$$\cos \theta = \frac{j'(j'+1) - L(L+1) - j(j+1)}{2jL}$$

Then in the limit of large L, j and j', corresponding to the classical vector model, we have

$$W(L\nu j'j; Lj) \rightarrow \frac{(-)^\nu P_\nu(\cos \theta)}{[(2L+1)(2j+1)]^{\frac{1}{2}}} \sim \tfrac{1}{2}(-)^\nu (Lj)^{-\frac{1}{2}} P_\nu(\cos \theta)$$

B2. Racah Coefficients with Repeated Parameters

For many purposes the Racah coefficient $W(jjLL; \nu j')$ is important; see B1 above.   For example, this coefficient occurs in the matrix elements of the interaction energy of a spin with its surroundings, section 29, and in angular correlation, sections 33 and 34.   The calculation of these specialized coefficients is based on a specialization of the general recurrence formula which all Racah coefficients satisfy.[4]   Defining $K_\nu(jj'L)$ by

$$W(jjLL; \nu j') = (-)^{j+L-\nu-j'} \left[\frac{(2j-\nu)!(2L-\nu)!}{(2j+\nu+1)!(2L+\nu+1)!}\right]^{\frac{1}{2}} K_\nu(jj'L)$$

[4] L. C. Biedenharn, *Oak Ridge Natl. Lab. Rept. 1098.*

The recurrence formula is

$$K_{\nu+1} = \frac{2\nu + 1}{\nu + 1} K_1 K_\nu - (2\nu + 1) K_\nu$$

$$- \frac{\nu}{\nu + 1} [(2j + 1)^2 - \nu^2][(2L + 1)^2 - \nu^2] K_{\nu-1}$$

Defining A by

$$A = j'(j' + 1) - j(j + 1) - L(L + 1)$$

and with

$$K_0 = 1 \quad \text{and} \quad K_1 = -2A$$

we find

$$K_2 = 6A(A + 1) - 8L(L + 1) j(j + 1)$$

$$K_3 = -20A^3 - 80A^2 + 16A$$

$$\times [3j(j + 1) L(L + 1) - j(j + 1) - L(L + 1) - 3]$$

$$+ 80j(j + 1) L(L + 1)$$

B3. Symmetry Relations

One often needs to apply several of the relations (6.10) in succession. In order to simplify this procedure, we present the relation between the Racah coefficients for all 24 possible combinations of the six arguments. The following permutation of $abcd$; ef are permissible without phase change:

$(badc; ef)$ $(cdab; ef)$ $(dcba; ef)$ $(acbd; fe)$ $(cadb; fe)$ $(bdac; fe)$ $(dbca; fe)$

The Racah coefficients of the following argument permutations give $(-)^{b+c-e-f} W(abcd; ef)$:

$(aefd; bc)$ $(eadf; bc)$ $(fdae; bc)$ $(dfea; bc)$ $(afed; cb)$ $(fade; cb)$ $(edaf; cb)$

$(defa; cb)$

The Racah coefficients of the following argument permutations give $(-)^{a+d-e-f} W(abcd; ef)$:

$(ebcf; ad)$ $(befc; ad)$ $(cfeb; ad)$ $(fcbe; ad)$ $(ecbf; da)$ $(cefb; da)$ $(bfec; da)$

$(fbce; da)$

B4. Tables of Simple Racah Coefficients

Tables I.3 and I.4 give $W(abcd; ef)$ for $e = \frac{1}{2}$ and $e = 1$.

Table I.3. $W(abcd; \tfrac{1}{2}f)$

	$a = b + \tfrac{1}{2}$	$a = b - \tfrac{1}{2}$
$c = d + \tfrac{1}{2}$	$(-)^{b+d-f}\left[\dfrac{(b+d+f+2)(b+d-f+1)}{(2b+1)(2b+2)(2d+1)(2d+2)}\right]^{\frac{1}{2}}$	$(-)^{b+d-f}\left[\dfrac{(f-b+d+1)(f+b-d)}{2b(2b+1)(2d+1)(2d+2)}\right]^{\frac{1}{2}}$
$c = d - \tfrac{1}{2}$	$(-)^{b+d-f}\left[\dfrac{(f+b-d+1)(f-b+d)}{(2b+1)(2b+2)2d(2d+1)}\right]^{\frac{1}{2}}$	$(-)^{b+d-f-1}\left[\dfrac{(b+d+f+1)(b+d-f)}{2b(2b+1)2c(2c+1)}\right]^{\frac{1}{2}}$

Table I.4. $W(abcd; 1f)$

$$c = d + 1$$

$a = b + 1$	$(-)^{b+d-f}\left[\dfrac{(f+b+d+3)(f+b+d+2)(-f+b+d+2)(-f+b+d+1)}{4(2b+3)(b+1)(2b+1)(2d+3)(d+1)(2d+1)}\right]^{\frac{1}{2}}$
$a = b$	$(-)^{b+d-f}\left[\dfrac{(f+b+d+2)(-f+b+d+1)(f-b+d+1)(f+b-d)}{4b(2b+1)(b+1)(2d+1)(d+1)(2d+3)}\right]^{\frac{1}{2}}$
$a = b - 1$	$(-)^{b+d-f}\left[\dfrac{(f+b-d)(f+b-d-1)(f-b+d+2)(f-b+d+1)}{4(2b+1)(2b-1)b(d+1)(2d+1)(2d+3)}\right]^{\frac{1}{2}}$

$$c = d$$

$a = b + 1$	$(-)^{b+d-f}\left[\dfrac{(f+b+d+2)(f+b-d+1)(b+d-f+1)(f-b+d)}{4(2b+1)(b+1)(2b+3)d(d+1)(2d+1)}\right]^{\frac{1}{2}}$
$a = b$	$(-)^{b+d-f}\dfrac{b(b+1)+d(d+1)-f(f+1)}{[4b(b+1)(2b+1)d(d+1)(2d+1)]^{\frac{1}{2}}}$
$a = b - 1$	$(-)^{b+d-f-1}\left[\dfrac{(b+d+f+1)(b+d-f)(f+b-d)(f-b+d+1)}{4(2b+1)b(2b-1)d(2d+1)(d+1)}\right]^{\frac{1}{2}}$

$$c = d - 1$$

$a = b + 1$	$(-)^{b+d-f}\left[\dfrac{(f-b+d)(f-b+d-1)(f+b-d+2)(f+b-d+1)}{4(2b+1)(b+1)(2b+3)(2d-1)d(2d+1)}\right]^{\frac{1}{2}}$
$a = b$	$(-)^{b+d-f-1}\left[\dfrac{(f+b+d+1)(f+b-d+1)(f+d-b)(b+d-f)}{4b(2b+1)(b+1)d(2d+1)(2d-1)}\right]^{\frac{1}{2}}$
$a = b - 1$	$(-)^{b+d-f}\left[\dfrac{(f+b+d+1)(f+b+d)(-f+b+d)(-f+b+d-1)}{4(2b+1)b(2b-1)(2d+1)d(2d-1)}\right]^{\frac{1}{2}}$

APPENDIX II. THE ROTATION MATRICES

Since the rotation matrices D^j play such an important role in the theory of angular momentum, it is desirable to present the derivation which leads to the explicit result (4.13). This is a more convenient procedure for the explicit definition of these matrices than that provided by the recurrence relation of section 15. The derivation [1] consists of two steps. In the first (A below) we show that there is a correspondence (isomorphism) between the 2-by-2 unitary matrices and the rotations in three-dimensional space. In the second (B) we obtain a representation of the unitary group in $2j+1$ dimensions. This is then to be identified with the $2j+1$-dimensional representation of the rotation group; see section 13.

A. We consider the unitary transformations on two variables λ_1 and λ_2:

$$\lambda_1' = a_{11}\lambda_1 + a_{12}\lambda_2$$

$$\lambda_2' = a_{21}\lambda_1 + a_{22}\lambda_2 \qquad (\text{II}.1)$$

It was shown in section 15 that the unitary matrix with elements a_{ij} depends on three parameters; cf. (4.50). We can write the equations (II.1) as

$$\lambda' = u\lambda$$

where

$$u = \begin{pmatrix} a & b \\ -b^* & a^* \end{pmatrix} \qquad (\text{II}.2)$$

with $|a|^2 + |b|^2 = 1$. Alternatively, in the notation of (4.50)

$$a = e^{i\xi}\cos\omega, \qquad b = e^{i\eta}\sin\omega \qquad (\text{II}.3)$$

so that the determinantal condition is automatically satisfied.

We consider the unitary transformation defined by u applied to a 2-by-2 matrix. Clearly the most general 2-by-2 matrix R can be ex-

[1] See E. P. Wigner, *Gruppentheorie*, pp. 168–180, Friedrich Vieweg und Sohn, Braunschweig, 1931. Also W. I. Smirnow, *Lehrgang der Höhern Mathematik*, pp. 207–213, Deutscher Verlag der Wissenschaften, Berlin, 1954.

pressed as a linear combination of the unit matrix I and the three Pauli spin matrices $\boldsymbol{\sigma}$. Thus,

$$R = R_0 I + x\sigma_x + y\sigma_y + z\sigma_z = \begin{pmatrix} R_0 + z & x - iy \\ x + iy & R_0 - z \end{pmatrix} \tag{II.4}$$

Then,

$$R_0 = \tfrac{1}{2}(R_{11} + R_{22})$$

$$z = \tfrac{1}{2}(R_{11} - R_{22})$$

$$x = \tfrac{1}{2}(R_{21} + R_{12})$$

$$y = \frac{1}{2i}(R_{21} - R_{12}) \tag{II.5}$$

The transformed matrix is

$$R' = uRu^{-1} \equiv R_0'I + x'\sigma_x + y'\sigma_y + z'\sigma_z \tag{II.6}$$

and clearly $R_0' = R_0$ which is the well-known result that the trace is unchanged by a unitary transformation. However, the quantities x', y', z' are linearly related to x, y, z, and we denote this by [cf. (4.37)]

$$\mathbf{r}' = M\mathbf{r}$$

with $\mathbf{r}'$ and $\mathbf{r}$ as collective (vector) symbols for the triads x', y', z' and x, y, z, respectively. With

$$u^{-1} = u^+ = \begin{pmatrix} a^* & -b \\ b^* & a \end{pmatrix} \tag{II.2'}$$

we find for the elements of M:

$$M_{xx} = \tfrac{1}{2}(a^2 - b^2 + a^{*2} - b^{*2}) = \cos^2 \omega \cos 2\xi - \sin^2 \omega \cos 2\eta$$

$$M_{yy} = \tfrac{1}{2}(a^2 + b^2 + a^{*2} + b^{*2}) = \cos^2 \omega \cos 2\xi + \sin^2 \omega \cos 2\eta$$

$$M_{zz} = |a|^2 - |b|^2 = \cos^2 \omega - \sin^2 \omega = \cos 2\omega$$

$$M_{xy} = \frac{i}{2}(a^{*2} - a^2 + b^{*2} - b^2) = \cos^2 \omega \sin 2\xi + \sin^2 \omega \sin 2\eta$$

$$M_{yx} = \frac{i}{2}(a^2 - a^{*2} + b^{*2} - b^2) = -\cos^2 \omega \sin 2\xi + \sin^2 \omega \sin 2\eta$$

$$M_{yz} = i(a^*b^* - ab) = \sin 2\omega \sin (\xi + \eta)$$

$$M_{zy} = i(a^*b - ab^*) = \sin 2\omega \sin (\xi - \eta)$$

$$M_{zx} = ab^* + ba^* = \sin 2\omega \cos (\xi - \eta)$$

$$M_{xz} = -(a^*b^* + ab) = -\sin 2\omega \cos (\xi + \eta) \tag{II.7}$$

Since the determinant of a matrix is unchanged by the unitary transformation, it follows that

$$\det R = \det R'$$

or

$$x^2 + y^2 + z^2 = x'^2 + y'^2 + z'^2 \tag{II.8}$$

Therefore the matrix M corresponds to a rotation. The possibility of a reflection is excluded because by setting $a^2 = 1$ and $b = 0$ we find $\det M = 1$, and the value of this determinant is a continuous function of the parameters on which it depends. Therefore it can never be -1. The relation to the Euler angles is clear since by setting $\omega = 0$ or $b = 0$ we get

$$M = \begin{bmatrix} \cos 2\xi & \sin 2\xi & 0 \\ -\sin 2\xi & \cos 2\xi & 0 \\ 0 & 0 & 1 \end{bmatrix} \tag{II.7a}$$

and by setting $\xi = \eta = 0$ we find

$$M = \begin{bmatrix} \cos 2\omega & 0 & -\sin 2\omega \\ 0 & 1 & 0 \\ \sin 2\omega & 0 & \cos 2\omega \end{bmatrix} \tag{II.7b}$$

which correspond to rotations about the z-axis through 2ξ and about the y-axis through 2ω, respectively.

As a check on these results we observe that the matrix elements of M can be expressed quite generally by

$$M_{jk} = \tfrac{1}{2} \operatorname{trace} \sigma_j u \, \sigma_k u^{-1}$$

where we have used Latin indices in place of x, y, z. It is clear that the inverse matrix M^{-1} is obtained by transposing M (this matrix is real, of course), and this corresponds to interchanging u and u^{-1}. From (II.2) and (II.2$'$) it follows that this is equivalent to the replacements $a \rightarrow a^*$ and $b \rightarrow -b$ or $\xi \rightarrow -\xi$ and $\omega \rightarrow -\omega$. This conclusion is in agreement with (II.7) as well as (II.7a and b).

It is also evident that for each and every u there is a single rotation and a single matrix M. The identity $u = 1$ corresponds to the identity rotation (the 3-by-3 identity matrix). Also, if M_1 results from u_1 and M_2 from u_2, then $M_2 M_1$ results from $u_2 u_1$. This is seen as follows: First we recognize that in formal terms the phrase "M_1 results from u_1" means that, with $R = R_0 + \mathbf{r} \cdot \boldsymbol{\sigma}$, the transformation $R' = u_1 R u_1^{-1} =$

$R_0 + \mathbf{r}' \cdot \boldsymbol{\sigma}$ defines a matrix M_1 such that $\mathbf{r}' = M_1\mathbf{r}$. Then, applying two successive transformations,

$$R'' = u_2 R' u_2^{-1} = u_2 u_1 R u_1^{-1} u_2^{-1} = (u_2 u_1) R (u_2 u_1)^{-1}$$

Leaving aside R_0 as irrelevant, we can write $R = \mathbf{r} \cdot \boldsymbol{\sigma}$ and

$$R' = \mathbf{r}' \cdot \boldsymbol{\sigma} = M_1 \mathbf{r} \cdot \boldsymbol{\sigma}$$

$$R'' = \mathbf{r}'' \cdot \boldsymbol{\sigma} = M_2 \mathbf{r}' \cdot \boldsymbol{\sigma} = M_2 M_1 \mathbf{r} \cdot \boldsymbol{\sigma}$$

$$= M_2 M_1 R$$

Therefore the multiplication table of the 2-by-2 unitary matrices has precisely the form of the multiplication table of the M-matrices. It follows then that the u- and M-matrices are representations of one and the same group—the rotation group (see end of section 13).

The correspondence (isomorphism) is not one-to-one. In fact, the results (II.7) show that we could change the sign of both a and b without changing M. This, in fact, is the extent of the ambiguity. Both u and $-u$ give the same M, i.e., the same rotation. It is sufficient to consider $M = I$. Then by (II.7) it is necessary and sufficient that ω and ξ be integral multiples of π, and η may be arbitrary. This gives $u = \pm I$. The rotation group is therefore the smaller, having one element for every two elements in the two-dimensional unitary group. This could have been anticipated from the result that the Euler angles are twice the angles that appear in the unitary matrices u, so that a change of π in these angles makes a change of 2π in the rotation angles and, consequently, leads to the same rotation.

B. The second part of the argument is concerned with the construction of a set of matrices A^j, of dimensionality $2j+1$ which constitute a representation of the unitary group. Thus for every u there exists a matrix A^j ($j = 0, \frac{1}{2}, 1, \cdots$) with the same group properties (same multiplication table). Of course, $A^{\frac{1}{2}} = u$, and so this case is trivial. Even simpler is the case $j = 0$ where $A^0 = I = 1$.

The problem is to define $2j+1$ functions of λ_1 and λ_2 which will undergo unitary transformation when λ_1 and λ_2 are transformed as in (II.1). It is clear that the monomials

$$\lambda_1^{j+m} \lambda_2^{j-m}, \qquad -j \leq m \leq j$$

will undergo linear transformations. In order that these transformations be unitary as well, we adjoin a normalization constant N_{jm} and consider

$$\Lambda_{jm} = N_{jm} \lambda_1^{j+m} \lambda_2^{j-m} \tag{II.9a}$$

Then

$$\Lambda'_{jm} = N_{jm}\lambda_1'^{j+m}\lambda_2'^{j-m} \tag{II.9b}$$

where λ_1', λ_2 are given by (II.1) with $a_{11} = a_{22}^* = a$, $a_{12} = -a_{21}^* = b$. The normalization constant is fixed by the condition

$$\sum_{m=-j}^{j} \Lambda'_{jm}\Lambda'^{*}_{jm} = \sum_{m=-j}^{j} \Lambda_{jm}\Lambda^{*}_{jm} \tag{II.10}$$

But

$$\sum_m |\Lambda'_{jm}|^2 = \sum_m |N_{jm}|^2 (\lambda_1'\lambda_1'^*)^{j+m}(\lambda_2'\lambda_2'^*)^{j-m} \tag{II.11}$$

and, since

$$|\lambda_1|^2 + |\lambda_2|^2 = |\lambda_1'|^2 + |\lambda_2'|^2$$

is the only invariant of (II.1), the unitary property is assured if $|N_{jm}|^2$ is such that the sum in (II.11) is a function of $|\lambda_1'|^2 + |\lambda_2'|^2$. This can be the case only if

$$|N_{jm}|^2 \sim [(j+m)!(j-m)!]^{-1}$$

In that case, taking the proportionality constant equal to 1,

$$\sum_m |\Lambda'_{jm}|^2 = \frac{1}{(2j)!}(|\lambda_1'|^2 + |\lambda_2'|^2)^{2j} = \frac{1}{(2j)!}(|\lambda_1|^2 + |\lambda_2|^2)^{2j}$$

$$= \sum_m |\Lambda_{jm}|^2$$

by the binomial theorem. The phase in Λ_{jm} is irrelevant for our purposes, and we set

$$\Lambda_{jm} = \frac{\lambda_1^{j+m}\lambda_2^{j-m}}{[(j+m)!(j-m)!]^{\frac{1}{2}}} \tag{II.12}$$

Introducing the transformation (II.1), we have

$$\Lambda'_{jm} \equiv \sum_{m'} A^j_{m'm}\Lambda_{jm'}$$

$$= \sum_s \sum_{s'} (-)^{j-m-s'} \frac{[(j+m)!(j-m)!]^{\frac{1}{2}}}{s!s'!(j+m-s)!(j-m-s')!}$$

$$\times a^{j+m-s}a^{*s'}b^{*j-m-s'}b^s\lambda_1^{2j-s-s'}\lambda_2^{s+s'} \tag{II.13}$$

If we remember that $1/n! = 0$ for $n < 0$, it is actually unnecessary to specify the limits on the sums in (II.13) and in the sums that follow. The summation indices can actually go from $-\infty$ to ∞ since all extra terms thereby introduced actually have a vanishing coefficient. In (II.13) we set $m' = j - s - s'$ and obtain

$$\Lambda'_{jm} = \sum_s \sum_{m'} (-)^{m'+s-m} \frac{[(j + m)!(j - m)!]^{\frac{1}{2}}}{s!(j - s - m')!(j + m - s)!(m' + s - m)!}$$
$$\times a^{j+m-s}a^{*j-m'-s}b^{*m'+s-m}b^s \lambda_1^{j+m'}\lambda_2^{j-m'} \quad (II.14)$$

Using (II.12), this becomes

$$\Lambda'_{jm} = \sum_{m'} \sum_s (-)^{m'+s-m} \frac{[(j + m)!(j - m)!(j + m')(j - m')!]^{\frac{1}{2}}}{s!(j - s - m')!(j + m - s)!(m' + s - m)!}$$
$$\times a^{j+m-s}a^{*j-m'-s}b^{*m'+s-m}b^s \Lambda_{jm'}$$

The transformation matrix is therefore

$$A^j_{m'm} = (-)^{m'-m} \sum_s (-)^s \frac{[(j + m)!(j - m)!(j + m')!(j - m')!]^{\frac{1}{2}}}{s!(j - s - m')!(j + m - s)!(m' + s - m)!}$$
$$\times a^{j+m-s}a^{*j-m'-s}b^{*m'+s-m}b^s \quad (II.15)$$

and

$$-j \leq m \leq j, \qquad -j \leq m' \leq j$$

so that the A-matrices are $2j+1$-dimensional. As a check we observe that for the identity transformation $a = 1$, $b = 0$ the factor b^s requires $s = 0$ and the factor $b^{*m'+s-m}$ requires $m' = m$. Hence, we find

$$A^j_{m'm} = \delta_{m'm}$$

as expected.

For each and every two dimensional unitary matrix u there is one and only one $2j+1$-dimensional matrix A, and its elements are given by (II.15). If two matrices A and B of the latter class are determined by the matrices u_1 and u_2, then AB is determined in the same sense by u_1u_2. The $2j+1$-dimensional set constitutes a representation of the unitary group. It is evident that the correspondence of the A- and u-matrices is not always one-to-one. If a and b change sign, all elements of A are multiplied by a factor $(-)^{2j}$ which is 1 for integer j and -1 for half-integer j.

Recalling the results of part A, we may conclude that for integral j there is one matrix A for each rotation, since the ambiguity in u is of no relevance so far as the definition of the rotation is concerned. Therefore, the matrices A^j are in one-to-one correspondence with the rotations, in three-dimensional space, defined by three Euler angles α, β, γ. For half-integer j there are two matrices A^j and $-A^j$ which correspond to u and

$-u$, respectively, and therefore to the same transformation matrix M. Moreover, the matrix M defines the Euler angles within trivial additive angles equal to a multiple of 2π. Therefore, the matrices A^j define a two-to-one correspondence with the group of rotations. These are just the properties required of the rotation matrices D^j. In identifying these two sets of matrices, it is clear that we can write

$$D^j = UA^jU^{-1}$$

with some definition of a and b. Of course, U is unitary. With the choice [2] (cf. II.3 and section 15),

$$a = e^{-i\alpha/2}\cos\frac{\beta}{2}e^{-i\gamma/2}$$

$$b = e^{i\alpha/2}\sin\frac{\beta}{2}e^{-i\gamma/2} \tag{II.16}$$

we find $U = I$ and $D^j = A^j$. Thus,

$$
\begin{aligned}
&D^j_{m'm}(\alpha\beta\gamma) \\
&= e^{-im'\alpha}\,e^{-im\gamma}\sum_s \frac{(-)^s[(j+m)!(j-m)!(j+m')!(j-m')!]^{\frac{1}{2}}}{s!(j-s-m')!(j+m-s)!(m'+s-m)!} \\
&\quad\times\left(\cos\frac{\beta}{2}\right)^{2j+m-m'-2s}\left(-\sin\frac{\beta}{2}\right)^{m'-m+2s} \tag{II.17}
\end{aligned}
$$

in agreement with (4.12) and (4.13).

[2] See equation (4.50). It should be recalled that the first row in $d^{\frac{1}{2}}$ corresponds to $m' = \frac{1}{2}$ and the first column to $m = \frac{1}{2}$. Wigner, reference 1, numbers the rows and columns with m and m' reversed in sign. This is equivalent to choosing $U_{m'm} = \delta_{m',-m}$; that is, U is a matrix with 1 in all the elements of the diagonal running from upper right to lower left and zeros elsewhere.

APPENDIX III. THE SPHERICAL HARMONICS

It is our purpose here to discuss some of the properties of the orbital angular momentum eigenfunctions, the spherical harmonics on the unit sphere.

The spherical harmonics diagonalize $\mathbf{L}^2$ and L_z where $\mathbf{L} = -i(\mathbf{r} \times \nabla)$. Thus,

$$\mathbf{L}^2 Y_{lm} = l(l+1) Y_{lm} \qquad \text{(III.1)}$$

$$L_z Y_{lm} = m Y_{lm} \qquad \text{(III.2)}$$

They are of particular importance since every field may have orbital angular momentum whether the intrinsic spin exists or not.

Instead of using the Cartesian form

$$L_z = -i\left(x \frac{\partial}{\partial y} - y \frac{\partial}{\partial x} \right) \qquad \text{(c.p.)}$$

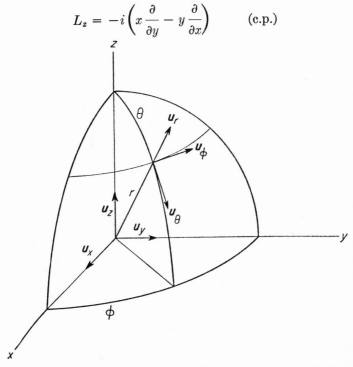

Fig. III. 1. Unit vectors in the Cartesian and spherical polar coordinate systems.

we shall write the orbital angular momentum operators L_x, L_y, and L_z in terms of differential operators on the unit sphere. Because $\mathbf{L} = -i(\mathbf{r} \times \nabla)$, the orbital angular momentum has no component in the radial direction. As we have already indicated in (III.2), we choose the z-axis as the direction of quantization. In spherical polar coordinates the coordinates of the point $\mathbf{r}$ are (r, θ, φ), where the colatitude θ and the azimuthal angle φ are shown in Fig. III.1. The Cartesian coordinates are $x = r \sin \theta \cos \varphi$, $y = r \sin \theta \sin \varphi$, and $z = r \cos \theta$. We shall indicate unit vectors in the two systems by the symbol $\mathbf{u}$ with an appropriate index to indicate its direction. Thus, the unit vectors along the x-, y-, and z-axes will be $\mathbf{u}_x$, $\mathbf{u}_y$ and $\mathbf{u}_z$, respectively. Along the r, θ, and φ directions they are $\mathbf{u}_r$, $\mathbf{u}_\theta$, and $\mathbf{u}_\varphi$, respectively. Now the gradient operator in spherical coordinates is

$$\nabla = \mathbf{u}_r \frac{\partial}{\partial r} + \mathbf{u}_\theta \frac{1}{r} \frac{\partial}{\partial \theta} + \mathbf{u}_\varphi \frac{1}{r \sin \theta} \frac{\partial}{\partial \varphi}$$

and $\mathbf{L}$ is then

$$\mathbf{L} = -i(\mathbf{r} \times \nabla) = -ir\mathbf{u}_r \times \left(\mathbf{u}_r \frac{\partial}{\partial r} + \mathbf{u}_\theta \frac{1}{r} \frac{\partial}{\partial \theta} + \mathbf{u}_\varphi \frac{1}{r \sin \theta} \frac{\partial}{\partial \varphi} \right)$$

Since $\mathbf{u}_r \times \mathbf{u}_r = 0$, $\mathbf{u}_r \times \mathbf{u}_\theta = \mathbf{u}_\varphi$, and $\mathbf{u}_r \times \mathbf{u}_\varphi = -\mathbf{u}_\theta$, $\mathbf{L}$ assumes the following form:

$$\mathbf{L} = -i \left(-\mathbf{u}_\theta \frac{1}{\sin \theta} \frac{\partial}{\partial \varphi} + \mathbf{u}_\varphi \frac{\partial}{\partial \theta} \right) \tag{III.3}$$

This expression shows explicitly how $\mathbf{L}$ depends on differential operators on the surface of a sphere, and, since the radial coordinate does not enter, we arbitrarily take the radius of this sphere to be unity and call it the unit sphere. To find the particular forms for L_x, L_y, and L_z we need the components of $\mathbf{u}_\theta$ and $\mathbf{u}_\varphi$ on the Cartesian basis:

$$\mathbf{u}_x \cdot \mathbf{u}_\theta = \cos \theta \cos \varphi \quad \mathbf{u}_y \cdot \mathbf{u}_\theta = \cos \theta \sin \varphi \quad \mathbf{u}_z \cdot \mathbf{u}_\theta = -\sin \theta$$

$$\mathbf{u}_x \cdot \mathbf{u}_\varphi = -\sin \varphi \quad \mathbf{u}_y \cdot \mathbf{u}_\varphi = \cos \varphi \quad \mathbf{u}_z \cdot \mathbf{u}_\varphi = 0$$

Then we get these expressions

$$L_x = i \left(\cot \theta \cos \varphi \frac{\partial}{\partial \varphi} + \sin \varphi \frac{\partial}{\partial \theta} \right) \tag{III.4a}$$

$$L_y = i \left(\cot \theta \sin \varphi \frac{\partial}{\partial \varphi} - \cos \varphi \frac{\partial}{\partial \theta} \right) \tag{III.4b}$$

$$L_z = -i \frac{\partial}{\partial \varphi} \tag{III.4c}$$

or more concisely

$$L_\pm = e^{\pm i\varphi}\left(i\cot\theta\,\frac{\partial}{\partial\varphi}\pm\frac{\partial}{\partial\theta}\right)$$ (III.5)

$$L_z = -i\,\frac{\partial}{\partial\varphi}$$ (III.6)

Substitution of (III.6) for L_z into (III.2) gives

$$\frac{\partial}{\partial\varphi}\,Y_{lm} = im\,Y_{lm}$$

This reveals that the φ dependence of the spherical harmonics is given by $e^{im\varphi}$:

$$Y_{lm} = e^{im\varphi}\,\mathcal{P}_{lm}(\theta)$$ (III.7)

where m must be an integer for Y_{lm} to be single-valued. With the variables separated, a standard procedure for finding $\mathcal{P}_{lm}$ is to operate on (III.7) with the operator for $\mathbf{L}^2$, which can be gotten from (III.4), and obtain a second-order differential equation

$$\left[\frac{1}{\sin\theta}\frac{d}{d\theta}\sin\theta\frac{d}{d\theta} + l(l+1) - \frac{m^2}{\sin^2\theta}\right]\mathcal{P}_{lm}(\theta) = 0$$ (III.8)

The standard power series solution of this equation is tedious, and instead we use a procedure based on properties of operators already introduced. The method is based on the use of the raising and lowering operators, which permit any spherical harmonic Y_{lm} to be found once a particular $Y_{lm'}$ is known, the order l being the same; see section 8.

The matrix elements of the raising and lowering $L_\pm = L_x \pm iL_y$ are found from (2.28), so that

$$L_\pm\,Y_{lm} = [(l\mp m)(l\pm m+1)]^{\frac{1}{2}}\,Y_{l,m\pm1}$$ (III.9)

For the particular cases $m = \pm l$,

$$L_+\,Y_{ll} = 0$$ (III.10a)

$$L_-\,Y_{l,-l} = 0$$ (III.10b)

since there are no eigenfunctions for $|m| > l$. Writing $Y_{l,\pm l} = e^{\pm il\varphi}\,\mathcal{P}_{l,\pm l}(\theta)$ and using (III.5) for $L_\pm$, this pair of equations becomes

$$\pm e^{i(l\pm1)\varphi}\left(\frac{d}{d\theta} - l\cot\theta\right)\mathcal{P}_{l,\pm l}(\theta) = 0$$

The exponential factor in this result does not vanish, of course, and so can be factored out to give the same simple differential equation for $\mathcal{P}_{ll}$ as for $\mathcal{P}_{l,-l}$:

$$\left(\frac{d}{d\theta} - l\cot\theta\right)\mathcal{P}_{l,\pm l}(\theta) = 0 \tag{III.11}$$

The solution of this first-order equation is, of course, far easier to obtain than the general solution of the second-order differential equation for $\mathcal{P}_{lm}$ in (III.8):

$$\mathcal{P}_{l,\pm l}(\theta) = C_{\pm}(\sin\theta)^l \tag{III.12}$$

Although the two functions have the same dependence on θ, they will have different constant multiplicative factors $C_{\pm}$. Thus, it is a simple matter to obtain the spherical harmonics $Y_{l,\pm l}$. All of the others will now be obtained from either of these two by application of the raising and lowering operators.

Let us decide to obtain all the spherical harmonics by successive application of the raising operator to $Y_{l,-l}$. A second and a third application of L_+ to (III.9) gives

$$L_+^2\, Y_{lm} = [(l-m)(l+m+1)(l-m-1)(l+m+2)]^{\frac{1}{2}}\, Y_{l,m+2}$$

$$= \left[\frac{(l-m)!(l+m+2)!}{(l-m-2)!(l+m)!}\right]^{\frac{1}{2}} Y_{lm+2}$$

$$L_+^3\, Y_m = \left[\frac{(l-m-2)(l-m)!(l+m+3)(l+m+2)!}{(l-m-2)!(l+m)!}\right]^{\frac{1}{2}} Y_{l,m+3}$$

$$= \left[\frac{(l-m)!}{(l-m-3)!}\frac{(l+m+3)!}{(l+m)!}\right]^{\frac{1}{2}} Y_{l,m+3}$$

It is easy to see that k applications of L_+ on Y_{lm} result in

$$L_+^k\, Y_{lm} = \left[\frac{(l-m)!(l+m+k)!}{(l-m-k)!(l+m)!}\right]^{\frac{1}{2}} Y_{l,m+k} \tag{III.13}$$

Finally, we can obtain Y_{lm} by $m+l$ applications of L_+ to $Y_{l,-l}$; i.e., we first set $m = -l$, and then let $k = m+l$:

$$Y_{lm} = \left[\frac{1}{(2l)!}\frac{(l-m)!}{(l+m)!}\right]^{\frac{1}{2}} L_+^{l+m}\, Y_{l,-l} \tag{III.14}$$

The right side is now evaluated by repeated use of the following identity:

$$L_{\pm}\, e^{i\mu\varphi} f(\theta) = \mp e^{i(\mu\pm 1)\varphi}(\sin\theta)^{1\pm\mu}\frac{d}{d(\cos\theta)}[(\sin\theta)^{\mp\mu} f(\theta)] \tag{III.15}$$

where $f(\theta)$ is any function.   Thus,

$$L_+^2 \, e^{i\mu\varphi} f(\theta) = (-)^2 \, e^{i(\mu+2)\varphi}(\sin\theta)^{2+\mu} \frac{d^2}{d\cos\theta^2} [(\sin\theta)^{-\mu} f(\theta)]$$

by using (III.15) with μ replaced by $\mu + 1$ and $f(\theta)$ replaced by $- (\sin\theta)^{1+\mu} \dfrac{d}{d(\cos\theta)} [(\sin\theta)^{-\mu} f(\theta)]$.   In general, we find that

$$L_+^k \, e^{i\mu\varphi} f(\theta) = (-)^k \, e^{i(\mu+k)\varphi}(\sin\theta)^{\mu+k} \frac{d^k}{d\cos\theta^k} [(\sin\theta)^{-\mu} f(\theta)] \quad \text{(III.16)}$$

Applying this result to (III.14), i.e. $\mu = -l$, $k = l + m$, and $f(\theta) = \wp_{l,-l}(\theta) = C_-(\sin\theta)^l$, gives this expression for Y_{lm}:

$$Y_{lm} = (-)^{l+m} B_l \left[\frac{(l-m)!}{(l+m)!} \right]^{\frac{1}{2}} e^{im\varphi}(\sin\theta)^m \left[\frac{d}{d(\cos\theta)} \right]^{l+m} (\sin\theta)^{2l}$$

$$\text{(III.17)}$$

The normalization constant C_- has been replaced by

$$B_l = C_-/[(2l)!]^{\frac{1}{2}}$$

Because B_l is independent of m, it can be determined by applying the orthonormality condition for $m = 0$:

$$Y_{l0} = (-)^l B_l \left(\frac{d}{d\cos\theta} \right)^l (\sin\theta)^{2l}$$

$$\int d\Omega (Y_{l0})^2 = 1$$

where $\int d\Omega$ indicates integration over the full solid angle.   At this point, let us set $x = \cos\theta$:

$$Y_{l0} = B_l \frac{d^l}{dx^l} (x^2 - 1)^l$$

and use the fact that the spherical harmonics for $m = 0$ must be proportional to the Legendre polynomials

$$Y_{l0} = A_l P_l \qquad\qquad \text{(III.18)}$$

Then the normalization condition is

$$1 = \int d\Omega (Y_{l0})^2 = A_l^2 \int d\Omega (P_l)^2 = 2\pi A_l^2 \int_{-1}^{1} dx (P_l)^2$$

since P_l is independent of φ. Using the orthogonality property of the Legendre polynomials,

$$\int_{-1}^{1} dx (P_l)^2 = \frac{2}{2l+1}$$

A_l is found to be

$$A_l = \epsilon_l \left[\frac{2l+1}{4\pi}\right]^{\frac{1}{2}}$$

where ϵ_l is a phase factor; i.e., $|\epsilon_l|^2 = 1$. The normalization constant B_l in (III.17) is now found by evaluating (III.18) at $x = 1$:

$$B_l \left[\frac{d^l}{dx^l} (x^2 - 1)^l\right]_{x=1} = \epsilon_l \left[\frac{2l+1}{4\pi}\right]^{\frac{1}{2}}$$

where we have followed the usual convention for the Legendre polynomials, namely, $P_l(1) = 1$. To evaluate the left side we introduce

$$u = x - 1$$

$$v = x + 1$$

and use Leibnitz's theorem,

$$\frac{d^n}{dx^n} A(x) B(x) = \sum_{\nu} \frac{n!}{(n-\nu)!\nu!} \frac{d^\nu A(x)}{dx^\nu} \frac{d^{n-\nu} B(x)}{dx^{n-\nu}} \qquad \text{(III.19)}$$

with $A(x) = u^l$ and $B(x) = v^l$. The only derivative of u that does not vanish is the lth; so

$$\left[\frac{d^l}{dx^l} (x^2 - 1)^l\right]_{x=1} = l![(x-1)^0 (x+1)^l]_{x=1} = 2^l l!$$

Therefore

$$B_l = \frac{\epsilon_l}{2^l l!} \left[\frac{2l+1}{4\pi}\right]^{\frac{1}{2}}$$

and adopting the usual convention, $\epsilon_l = +1$, the spherical harmonics are

$$Y_{lm} = \left[\frac{2l+1}{4\pi} \frac{(l-m)!}{(l+m)!}\right]^{\frac{1}{2}} \frac{1}{2^l l!} e^{im\varphi} (-\sin\theta)^m \left[\frac{d}{d(\cos\theta)}\right]^{l+m} (\cos^2\theta - 1)^l$$

$$\text{(III.20)}$$

The connection (III.18) between the Legendre polynomials and the Y_{l0} is now simply

$$Y_{l0}(\theta, \varphi) = \left[\frac{2l+1}{4\pi}\right]^{\frac{1}{2}} P_l(\cos\theta) \qquad (III.21)$$

As an example we write down the first-order $(l = 1)$ spherical harmonics:

$$Y_{1m} = \left(\frac{3}{4\pi}\right)^{\frac{1}{2}} \frac{1}{r} \begin{cases} -\dfrac{1}{\sqrt{2}}(x + iy) \\[2mm] z \\[2mm] \dfrac{1}{\sqrt{2}}(x - iy) \end{cases} \qquad (III.22)$$

The three combinations of the Cartesian coordinates $\mp(1/\sqrt{2})(x \pm iy)$ and z transform, therefore, like the spherical harmonics of order one.

According to (III.7) and (III.21),

$$Y_{lm} = e^{im\varphi}\, \mathcal{P}_{lm}(\theta)$$

where $\mathcal{P}_{lm}$, the unnormalized associated Legendre polynomial, is a real function. We shall now verify the important property of the spherical harmonics

$$Y_{lm}^*(\theta, \varphi) = (-)^m\, Y_{l,-m}(\theta, \varphi) \qquad (III.23)$$

This implies that

$$\mathcal{P}_{l,-m}(\theta) = (-)^m\, \mathcal{P}_{lm}(\theta) \qquad (III.24)$$

which we shall prove by direct substitution of

$$\mathcal{P}_{lm}(\theta) = \frac{1}{2^l l!}\left[\frac{2l+1}{4\pi}\frac{(l-m)!}{(l+m)!}\right]^{\frac{1}{2}}(-\sin\theta)^m\left[\frac{d}{d(\cos\theta)}\right]^{l+m}(\cos^2\theta - 1)^l$$

Canceling common factors, (III.24) gives

$$\frac{(l-m)!}{(l+m)!}(x^2-1)^m\frac{d^{l+m}}{dx^{l+m}}(x^2-1)^l = \frac{d^{l-m}}{dx^{l-m}}(x^2-1)^l$$

with $x = \cos\theta$. Once more we introduce $u = x - 1$ and $v = x + 1$ and use (III.19),

$$\frac{(l-m)!}{(l+m)!}u^m v^m \sum_\nu \frac{(l+m)!}{(l+m-\nu)!\nu!}\frac{l!}{(l-\nu)!}\frac{l!}{(\nu-m)!}u^{l-\nu}v^{\nu-m}$$

$$= \sum_\nu \frac{(l-m)!}{(l-m-\nu)!\nu!}\frac{l!}{(l-\nu)!}\frac{l!}{(\nu+m)!}u^{l-\nu}v^{\nu+m}$$

We have also used

$$\frac{d^n}{dx^n} u^l = \frac{l!}{(l-n)!} u^{l-n} \quad \text{and} \quad \frac{d^n}{dx^n} v^l = \frac{l!}{(l-n)!} v^{l-n}$$

which depends on the fact that the derivatives of u and v with respect to x are unity. The summation index ν assumes only those values for which all the factorial arguments are positive. The substitution $\nu' = \nu - m$ on the left side reduces the above equation to an identity:

$$\sum_{\nu'} \frac{(l-m)!}{(l-m-\nu')! \nu'!} \frac{l!}{(l-\nu')!} \frac{l!}{(\nu'+m)!} u^{l-\nu'} v^{\nu'+m}$$

$$= \sum_{\nu} \frac{(l-m)!}{(l-m-\nu)! \nu!} \frac{l!}{(l-\nu)!} \frac{l!}{(\nu+m)!} u^{l-\nu} v^{\nu+m}$$

thus verifying (III.23).

AUTHOR INDEX

SUBJECT INDEX